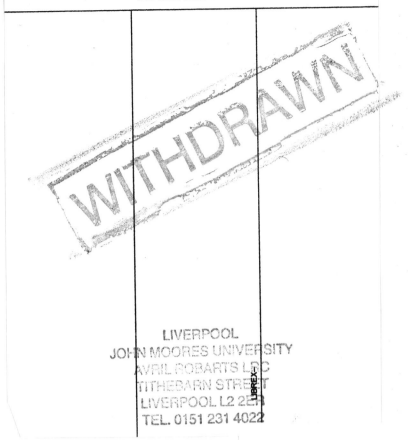

Risk Management in Projects

Project managers and professionals in construction, real estate, civil engineering and transport planning need to base their decisions on realistic information. For this, they require a sound understanding of risk and of people's attitudes towards it.

This second edition of John Raftery's successful and well-established text *Risk Analysis in Project Management* has expanded its focus to encompass the entire risk management process and to give a fuller presentation of how risk is perceived by the general public. This new material is covered by a complete reorganisation of the chapters on risk analysis and several new chapters on risk identification, risk perceptions, risk response, crisis and business continuity management and systems implementation.

A very straightforward and easy-to-read style has been retained, but there is an increased attention to the theory underlying the principles, and an expanded bibliography is given to guide an exploration of the subject in greater detail. The overall aim remains to de-mystify risk management by presenting the subject in simple and practical terms, free of technical jargon. Case studies are used extensively to enliven the text and to illustrate the concepts discussed.

Martin Loosemore is Professor and Associate Dean Research at the University of New South Wales, Sydney, Australia, and a Chartered Surveyor, Chartered Builder and Principal of a management consultancy LRC. He has published numerous books and articles in the areas of risk management, crisis management and human resource management.

John Raftery is Professor, Dean of the School of the Built Environment and Pro-Vice Chancellor (External Affairs), Oxford Brookes University, Oxford, UK. He has published many articles and books in the areas of risk management and construction economics and recently co-edited *The Construction Sector in the Asian Economies*.

Charlie Reilly is Deputy Managing Director, Multiplex Facilities Management Limited, and a Chartered Professional Engineer.

Dave Higgon is the Employee Relations Manager for Multiplex Constructions, NSW. He has over 30 years experience in the building and construction industry including time spent as a sub-contractor, trade union organiser and safety manager.

Also available from Taylor & Francis

Human Resource Management in Construction Projects: Strategic and Operational Approaches
Martin Loosemore, Andrew Dainty, Helen Lingard

Spon Press   Hb: 0-415-261635

The Construction Sector in the Asian Economies
John Raftery, Michael Anson, Y. H. Chiang

Spon Press   Hb: 0-415-286131

What Every Engineer Should Know About Risk Engineering & Management (Volume 36)
Martin L. Roush, John X. Wang

Spon Press   Hb: 0-824-793013

Effective Opportunity Management for Projects: Exploiting Positive Risk
Hillson

Hb: 0-824-748085

Information and ordering details

For price availability and ordering visit our website **www.tandf.co.uk/ builtenvironment.com**
Alternatively our books are available from all good bookshops.

Risk Management in Projects

Second edition

Martin Loosemore, John Raftery,
Charlie Reilly, Dave Higgon

Taylor & Francis
Taylor & Francis Group

LONDON AND NEW YORK

First published 1993
by E & FN Spon

Second edition published 2006
by Taylor & Francis
2 Park Square, Milton Park, Abingdon, Oxon OX14 4RN

Simultaneously published in the USA and Canada
by Taylor & Francis
270 Madison Ave, New York, NY 10016, USA

Taylor & Francis is an imprint of the Taylor & Francis Group

© 1993 John Raftery
© 2006 Martin Loosemore, John Raftery, Charlie Reilly, Dave Higgon

Typeset in Sabon by
Integra Software Services Pvt. Ltd, Pondicherry, India
Printed and bound in Great Britain by
TJ International, Padstow, Cornwall

British Library Cataloguing in Publication Data
A catalogue record for this book is available from the British Library

Library of Congress Cataloging in Publication Data
Risk management for projects / Martin Loosemore ... [*et al.*]. — 2nd ed.
    p.   cm.
    Includes bibliographical references and index.
    1. Construction industry—Risk management.   2. Building—Risk
    assessment.   3. Decision making.   4. Building—Safety measures.
    5. Emergency management.   I. Loosemore, Martin, 1962–
    TH438.R53 2005
    690'.068'4—dc22   2005014920

ISBN 10: 0–415–26055–8   ISBN 13: 9–78–0–415–26055–8 (hb)
ISBN 10: 0–415–26056–6   ISBN 13: 9–78–0–415–26056–5 (pb)

Contents

Tables

Figures

Foreword

As the secretary of Unions NSW I am acutely aware of the many benefits to be gained through a process of consultation that involves a genuine dialogue between employers and employees.

Too often the resolution of issues that arise in a workplace that may affect a project workforce and at times the wider community is seen as a management prerogative. This fact combined with an increasing absence of clear definition, clarity in performance standards and effective risk management has created some unfortunate consequences. Amongst these is the introduction of new jargon and an approach to dealing with problems that is often not understood by those at the receiving end of outcomes.

The authors of *Risk Management in Projects* present a persuasive argument, that for risk management to be ultimately successful it must be inclusive and reflect the legitimate interests of all the stakeholders. Those stakeholders include Unions, their members and their families.

This book, in addition to identifying the many advantages inherent in an inclusive approach to managing project risk, provides a valuable reference point for anyone involved in the areas of Health and Safety, the Environment and Workplace Relations.

I am aware of the authors' long standing involvement in the building and construction industry and share their commitment to making it an industry of first choice through the application of progressive employment practices, high standards of health and safety, environmental responsibility and leading edge performance in every area of its operations.

The authors have put together a plain English, easily accessible book which will provide invaluable assistance to those in the Building and Construction Industry seeking to manage project risks in a manner that gains the widest possible community support. In addition the book deserves to enjoy a more general readership in the fields of Human Resources and Occupational Health and Safety Management.

John Robertson, Secretary, Unions NSW

John Robertson was elected Secretary of Unions NSW (formerly the Labor Council New South Wales) on 26 April 2001. During his time at the Labor Council John's areas of responsibilities have included: the building and construction industry, electricity industry, breweries, local government, warehouse and distribution, oversight of public sector policy and negotiations, major disputes, the oil industry and Sydney Water. John is also on the Board of: Labor Media Pty Ltd, Getonboard, Labor Campaign, Building and Construction Industry Long Service Payments Committee, WorkCover New South Wales, Parramatta Stadium Trust, North South West Co-operative Housing Society Group Ltd, Homeseekers Co-operative Housing Society, City Central Co-operative Housing Society, Macquarie Co-operative Housing Society, NSW Co-operative Housing Society, Combined Unions Co-operative Housing Society.

Preface

In simple terms, risk management in projects is about proactively working with project stakeholders to minimise the risks and maximise the opportunities associated with project decisions. The aim is not to avoid risk but to make more informed decisions to ensure that project objectives are achieved and, ideally, exceeded. While the principles that underpin risk management are simple, the subject has become intimidating and impenetrable for many managers because of the highly mathematical, technical, academic and impractical style in which it is sometimes presented. Too many students are taught about risk in the abstract but not how to translate these ideas into practice when things go wrong.

Effective risk management demands a clear, straightforward and sensible approach. To this end, our aim is to demystify the subject by providing an accessible text which illustrates the practical application of recent risk management research and theory. We minimise mathematics and use simple, jargon-free language wherever possible. Our aim is not to propose one best approach to risk management but to bring professionals up-to-date with contemporary ideas, techniques and debates in the field which can add value to a project or a business. In doing this, we give equal attention to the management of opportunities as to the management of risks. We also pay particular attention to the social, ethical and psychological aspects of the risk management process. Our aim is to encourage managers to recognise and value the full diversity of interests in a project and to engender a sense of collective responsibility throughout the supply chain in order to manage effectively, both risks and opportunities. These features, and many others, distinguish this book from others available on risk management which tend to adopt a highly technical approach. Ultimately, our objective is to develop a maturity of judgement regarding the management of risks and opportunities, thereby improving levels of project performance for all project stakeholders and customers.

This book is based on a series of workshops, consultancies and research projects undertaken by the authors over the past 25 years, involving some of the largest international clients, consultants and contractors in Europe,

Asia, America and Australia. We frequently draw upon this experience to provide numerous practical examples and case studies of how research and theory can be of great value in managing major engineering, building and facilities management projects.

Risks and opportunities do not respect disciplinary boundaries and occur over the entire life cycle of a project. Therefore this book will be of interest to everyone involved in a construction supply chain (clients, public, statutory authorities, project managers, designers, engineers, surveyors, facilities managers, contractors, sub-contractors, suppliers and unions). It will also be of interest to those with responsibility for occupational health and safety, environmental management, public relations, insurance, contracts and legal issues, industrial relations, corporate governance and strategic planning. Finally, it will be of interest to academics and students of construction and engineering because it integrates theory and practice and presents case studies which report the results of new research.

Key features of this book:

- Written in a simple, jargon-free style.
- Practical, clear and easy to follow.
- Extensive use of real-life case studies and examples of principles in practice.
- Integrates risk management theory and practice.
- Based on recent research and best practice of risk management in other industries.
- Demonstrates how innovative companies can utilise contemporary risk management research to drive system and cultural reform.
- Focusses on opportunities as well as risks.

Acknowledgments

The authors wish to express their gratitude to the following people who: provided valuable feedback on the various drafts of this book; provided permission for access to data; helped develop Multiplex Facilities Management's new Risk and Opportunity Management System and/or; undertook research into the case studies.

Craig Barass, Multiplex Facilities Management Pty Ltd
Mario Barrios, Construction, Forestry, Mining and Energy Union, Australia
Allen Barry, Barrington Security
Alan Colwill, Multiplex Facilities Management Pty Ltd
David Elton, Multiplex Facilities Management Pty Ltd
Graham Ferguson, Multiplex Facilities Management Pty Ltd
Ian Gavin, Independent Consultant, Sydney, Australia
Richard Hyde, Westpac Bank, Sydney, Australia
Paul Jagla, Multiplex Construction, Victoria
Malcolm Jones, Defence Maintenance Management Pty Ltd
James Kell, Kell and Rigby
Amy Lam, University of New South Wales
Renita Lee, Multiplex Facilities Management Pty Ltd
Jim McGreevy, Multiplex Construction, New South Wales
Rob MacPherson, Defence Maintenance Management Pty Ltd
Jon McCormick, Multiplex Facilities Management Pty Ltd
Denny McGeorge, University of New South Wales
Andrew Merriel, Multiplex Construction, New South Wales
Angus Morton, Multiplex Construction, Queensland
Angie Ng, University of New South Wales
Frank Princi, Defence Maintenance Management Pty Ltd
Vanessa Sammons, University of New South Wales
Colin Smith, Multiplex Facilities Management Pty Ltd

Chris Stannage, Multiplex Construction, New South Wales
Ayu Suartika, University of New South Wales, Sydney
Stephen Thomas, Multiplex Facilities Management Pty Ltd
Mark Trevithic, Multiplex Facilities Management Pty Ltd.

Chapter 1

Risk and uncertainty in projects

> This is not a speculative game at all. Our objective is not to avoid risk but to recognise it, price it and sell it.
> – Tony Ryan, Chairman of Guinness Peat Aviation Ltd.

1.1 Introduction

Risk is a complex phenomenon that has physical, monetary, cultural and social dimensions. The consequences of risk events go well beyond the direct physical harm to financial or physical assets, people or ecosystems to effect the way a society operates and people think. Our experience of risk is not only shaped by the extent of potential harm but also by the way in which we interpret or "filter" information about it. This is vividly illustrated in the construction and engineering industries where prominent examples of poor service delivery, poor safety and environmental damage have created negative public perceptions and seemingly irrational and unjustified public reactions to even the most innocuous development proposals. Of course, to those holding these perceptions, this behaviour is not irrational at all. Nevertheless, these negative perceptions are too often ignored and indeed exacerbated by insufficient community consultation which excludes the different interests that have a legitimate role to play in the assessment and control of development risks.

The above problems reflect a preoccupation with the technical aspects of risk and a failure to consider risk in its full social and cultural complexity. They also reflect a perception that some industries are a unique and special case, demanding a different management approach to other industries. Arguments put forward to support this perception include:

- There are special problems in construction.
- It is not like other industries.
- It is a special case and should be treated differently.
- Every project is different.

- The future cannot be forecast in construction.
- Construction is a high-risk business.

These arguments are often accompanied by suggestions that different rules and regulations (covering service quality and reliability, product quality, working conditions and payment) should apply to construction, compared to other industries. We do not agree with most of this. Of course, many industries have their own special characteristics but few are so special that they merit special rules. In fact, viewed systematically, most construction projects are like any other project. There is a start and finish date; there are defined and limited resources; plans, estimates and time schedules have to be made in advance on the basis of limited information; they are procured by teams of people and firms drawn together for that particular project; these teams often change throughout the project; they require contractors to marshal labour equipment and materials and components to a specific site; and judgements about the future have to be made and built into plans and forecasts. Furthermore, construction projects in general tend to exhibit lower levels of complexity when compared to projects in other high-risk industries like aerospace, chemical engineering, oil and gas exploration, nuclear power, defence or computer software. These industries tend to deal with more complex projects, larger budgets and more dynamic and hostile environments. The problem in the construction industry has less to do with its dynamic environment but more to do with difficulties in identifying, assessing and managing the risks and opportunities posed by projects. There is little doubt that viewing the industry in terms of its special difficulties is rather inward looking and unhelpful. It also produces excessive concern with the negative aspects of risk and, in turn, to resistance to new ideas which could drive innovation and performance improvement.

In construction, as in all industries, it is necessary for firms to strike a balance between the avoidance of all risks on the one hand, and rash, risk-seeking behaviour on the other. The challenge is not to avoid risk, but to take calculated risks by recognising and managing them effectively. Indeed, there is a large, profitable and untapped market in the construction industry for firms which can manage their risks effectively. Ultimately, organisations make money by charging customers a fee to manage their risks. Firms which can manage effectively more risks than their competitors can make more money. For example, a building owner will pay a facility management company a fee to take and manage the many risks which can affect the efficient operation of a building in use. These risks include ongoing maintenance of building fabric and mechanical and electrical services, security, parking, cleaning, catering etc. The fee charged for managing these risks will be determined by the level of confidence to manage them effectively. The more confident a company is in its risk management systems, the more likely it is to secure work at a lower price than its competitors. More importantly, the more likely it is to be able to turn these risks into opportunities to make profit. The rewards for managing risks effectively can be enormous and are not confined to financial alone. Risk has about it, an aura of achievement and those who deal successfully with risky situations are held in high regard. Risk management is as much about developing a positive reputation in the long term as it is about making money in the short term.

The purpose of this chapter is to introduce some basic notions of risk, to describe the emergence of risk management as a preoccupation in business and to critically analyse traditional approaches to risk management in construction and facilities management. We argue for a life cycle approach to risk, which emphasises the collective responsibility and interdependence of all involved in each stage of the building procurement process. This contrasts with traditional approaches to risk management which tend to treat design, construction and facilities management as distinct and independent phases of activity. This has never been appropriate but with the development of Build Operate Transfer (BOT), Private Public Partnership (PPP) and Private Finance Initiative (PFI) procurement systems, the important interrelationships between the risks generated in these three phases of activity are being brought more sharply into focus. We see this as a very positive trend.

We also propose a more socially sensitive approach to risk management, which considers the needs of all stakeholders and recognises how the activities of the construction and facilities management industries affect society. This is important in all projects but especially in large projects which arouse special community interest and expectation. In particular, because many large projects are now procured via public–private partnerships, there is an increasing pubic expectation that shared project ownership means shared project information. Indeed, Sharpe (2004), from the Victorian Department of Treasury and Finance in Australia, which annually procures over $1.8 billion of public infrastructure, noted that the political and social context in which PPP projects are undertaken requires that public consultation be fully integrated into the planning process. He recognised that while PPP projects might offer creative solutions to public infrastructure needs, at the core is a complex web of relationships among bureaucrats, politicians, media, employees, general public (local, national and sometimes global), labour and special interest groups. Irrespective of any ideological preferences that governments of the day may promote, these interest groups invariably have high expectations in relation to the management of issues such as the environment, health and safety, industrial relations and access and equity. As Sharpe states, "Any PPP lives or dies on its reputation with these people" (Sharpe 2004: 8).

1.2   Background – A new era of risk

Risk came into the foreground of business literature during the last two decades of the twentieth century. Risk management, once taught only in engineering and finance courses, is now a subject on almost every undergraduate and postgraduate degree which leads to a career in business. The general public's awareness of risk has also been heightened. People are better educated and informed than at any time in history and are making ever more informed and conscious decisions about the social and environmental risks associated with the products they buy. For instance, in the 1980s many millions of people boycotted South African products in protest against apartheid, helping to bring down the South African government. Furthermore, in the 1990s a series of health scares and cover-ups associated with British beef, led to purchasing boycotts which caused the decimation of that industry. Beck (1992) argued that these events reflect an increasingly paranoid "risk society" where even an unsubstantiated claim about a threat to public health is likely to be taken seriously. This is being perpetuated by incessant media reporting about the risks to public health posed by terrorism, nuclear power, cloning, genetic engineering, pesticides, pollution, global warming, smoking, the latest computer virus and the spread of AIDS – to name but a few. The activities of many industries like construction have also come under the spotlight, presenting new challenges for managers which are not well understood but which are likely to grow.

In addition to the public's increasing preoccupation with risk, recent historical events have marked a turning point in the way that many companies manage their risks. Over the last few years atrocities such as September 11, 2001, the Bali bombing and the London bombings, health scares such as the SARS and Asian Bird Flu epidemics, the war in Iraq, high-profile corporate collapses such as HIH in Australia, ENRON in the US and AMP in the UK, the subsequent turmoil in the insurance industry and the recent financial reporting debacles in the USA, have had both tragic consequences and deeply unsettling side effects on both individuals and businesses. For example, business has experienced spiralling insurance premiums (over 100 per cent in some cases), reductions in insurance coverage and stricter claims regimes. Over the last three years, property owners' insurance costs have risen from around 5 per cent of costs to between 10 and 20 per cent of costs. Furthermore, before September 11, 2001 average insurance cover fell short of actual losses by up to 40 per cent but since then this gap has been rising dramatically, leaving companies to make up the difference (Hemsley 2002, Anonymous 2003). Indeed, for many organisations only 20 per cent of risks are now insurable, although insurance costs have increased to typically over 50 per cent of risk management costs (Fenton-Jones 2003). Those companies that traditionally relied on insurance as a mechanism for managing their risks are now counting the costs and realising that risk management is no longer the sole province of insurance and finance companies. It is now everyone's responsibility. While in the past, an effective risk management system may have been a useful marketing feature, it is now a basic necessity for every organisation. Insurance can be an inefficient way to manage many business risks and business clients in all industries are increasingly insisting on demonstrable capabilities in risk management as a critical selection criterion in tendering processes. With recent increases in risk-related legislation and a more informed public, the risk management practices of companies are under increasing scrutiny and the penalties for non-compliance are becoming more severe. In this environment, companies that ignore, or are slow to respond to their business risks will, in turn, eventually find that their reputation and survival is at risk. This is well illustrated in the case of James Hardy, a building product manufacturer, which has is currently experiencing the wrath of the public, unions, media, regulatory bodies and government for its handling of compensation for victims of its asbestos products. A special government commission established to investigate the company in response to public outcry found that the company had been aware of the dangers of asbestos since the 1930s but had continued to manufacture it until 1987, despite a mounting death toll from its products. It also found that the company had deliberately schemed to avoid its $1.5 billion compensation liabilities via a tortuous series of corporate transactions which included shifting its headquarters offshore to Holland (West 2004). The result of the debacle is that company directors currently face the possibility of serious criminal prosecutions for fraud and negligence. As West (2004: 4) points out "if the stakes are high enough, no amount of executive pettifogging and elite legal defence can withstand community demands for corporate responsibility".

While reacting to increasing insurance premiums and stricter compliance regimes are good reasons for investing in better risk management practices, there are far more positive reasons for doing so. For example, effective risk management provides a better basis for decision-making at strategic, tactical and operational levels by providing a robust, logical, systematic and transparent auditable process that uses best available experience to provide a clear understanding of potential risk profiles and options for dealing with them. Other important benefits include: better corporate reporting; better use of human resource expertise; increased engagement with stakeholders; less adverse publicity; a better basis for negotiations; reduced finance costs; increased reliability and quality of services and products; lessons and feedback to improve future business activities; reduced claims and legal costs; better change management; enhanced morale; reduced levels of conflict and stress; and ultimately enhanced competitive advantage. Indeed, the success of projects within an environment of increasing risk transfer from the public to private sector and ever tightening resource and time constraints depends more than ever on effective risk management processes. In this context, a single averted risk can repay investments in risk management many times over, whereas a single unidentified risk can cripple a project or a business.

1.3   Risk in the construction process

Those involved in the procurement of buildings are not immune to the trends discussed above. The events on and after September 11th 2001 have recalibrated the risks associated with financing, designing, constructing and managing facilities. Many of the new security threats people face today, revolve around the apparent exposure and vulnerability of buildings and infrastructure to such attacks. For instance, facilities managers should be aware that with new high efficiency ventilation and water supply systems, very large buildings containing many thousands of people could be completely contaminated by a biological attack in a few minutes (CIS 2002, Perinotto 2002). Indeed, so significant is this risk that the US Government has established an "Immune Building Programme" to develop intelligent building systems that can detect and react to terrorist risks automatically. Clearly, it is now more important than ever to think about sharing design information during the planning, construction and operational phases of a facility's life. And there is plenty of scope to improve practices. According to ASCE (2002), most US companies spend more on coffee than on information security. This is a major problem, especially with the trend towards working with international project teams where design, construction and facilities management information is frequently exchanged across the airwaves.

The problems of managing risks during the construction stage of the building procurement have also been exacerbated by the heightened public risk perceptions discussed in the previous section. To the average person, construction is an industry which is male dominated, dirty, dangerous, unattractive, environmentally destructive and unprofessional (Loosemore *et al.* 2004). The evidence for this view is overwhelming. For example, in Europe, construction accounts for over 15 per cent of workplace accidents despite representing less than 10 per cent of the working population. Similarly, in the US, construction accounts for 19 per cent of workplace fatalities, whilst accounting for only 6 per cent of the labour force. In Australia, the rate of compensated injuries and disease in the construction industry is 37.4 per 1000 workers, which is 63 per cent higher than the all-industry average of 22.9. The potential environmental, cultural and social impact of the industry is equally startling. For example, construction contributes a disproportionately large amount to landfill waste (Teo 2001) and one only has to look at mega projects such as the Three Gorges Dam in China to see its potential environmental impact. This project, which will supply 10 per cent of China's future electricity demand and improve economic conditions and navigation capabilities in the upper reaches of the Yangtze, has submerged over 8000 ancient archaeological sites and displaced over 1.9 million people causing untold social and environmental problems. There has been substantial social unrest resulting from the mass relocation of people and the resettlement programme has been associated with allegations of corruption, mismanagement and shortages in land and compensation (Salazar 2000). The ecological impact appears to have been equally enormous. Salazar reports that thousands of animals have been killed, and pristine natural habitats destroyed. There have also been increases in deforestation associated with new settlements and a 30 per cent increase in levels of silt carried by the river which in turn, has increased the threat of flooding and fresh water contamination upstream in the city of Chongqing (population over 30 million).

  While the vast majority of firms in the construction industry are not involved in projects of this magnitude, the collective social and environmental impact of the many thousands of small projects is much larger. Every week the newspapers, construction journals and magazines are littered with literally hundreds of examples of much smaller corporate crises and project disasters, which pollute the environment or cause suffering, death and financial loss, provoking condemnation from politicians, regulatory bodies, shareholders and the public. Of course, the many successful projects go unreported and hindsight is a wonderful management tool. Nevertheless, there is a lesson to be learnt because in the majority of cases, the companies involved in these incidents did not have an effective risk management system in place or did not implement it effectively. Many failed projects can be traced back to either a lack of discipline or lack of process in risk management, although it is always difficult to identify one particular party who is at fault.

  Research indicates that the entire supply chain has a responsibility to work collectively to manage project risks and that clients, contractors, consultants, suppliers and sub-contractors all contribute to this problem (Kumeraswamy *et al.* 2000; Loosemore 2000). For example, a common difficulty for many project teams is that there is too little time during the bidding phase to undertake all the necessary studies to identify, assess and control risks throughout the life cycle of a project. Furthermore, many important people such as facilities managers who can advise on operating risks are excluded from the tendering process. Indeed, when competitive tendering is used, it often becomes inefficient and irrational for bidders to explore risks with any rigour, since there is a good chance that if they do, their bids will not succeed. Disturbingly, even when the tendering environment is conducive to effective risk management, many companies do not have reliable risk-related data, well-developed risk management procedures and systems or the management discipline to apply them rigorously. Evidence of this is contained in the many project programmes, estimates and tenders that fail to reflect the variety of risks that can affect project outcomes. Due to the pressures of tendering and/or a lack of understanding of risk, too many schedulers and estimators make their forecasts and projections match targets by "cooking" the numbers. The consequence on most projects is a stark difference between final estimates and programmes and the original ones, although these comparisons are rarely made and analysed in practice. The result, when the project commences, is a continual stream of unexpected problems which have to be managed reactively. These problems can be addressed with a better understanding of the risk management process and a useful starting point is to understand of some basic terminology.

1.4   Risk management terminology

Although there has undoubtedly been a growth in risk management since the first edition of this book, the subject can still be intimidating to many students and managers. This is unnecessary because the ideas and concepts underlying risk management are very simple, common sense and used by most people on a day-to-day basis. Below, we discuss the most common terms used.

1.4.1   Risk

To most managers, risk is concerned with unpredictable events that might occur in the future whose exact likelihood and outcome is uncertain but could potentially affect their interests/objectives in some way (normally adversely). This implies a number of things.

First, there is an emphasis on the uncertainty. This is the absence of information about future events which makes them unpredictable. A certain future event which is predictable is not a risk but is a problem which needs resolving. This distinction between predictability and unpredictability is important because the future is largely unknown and most business decision-making takes place on the basis of expectations about the future. To reduce this uncertainty means gathering more information about a future event. However, managers often have little information about the future and in most instances have to base their forward judgements on past information. While this is a necessity in many decisions, it is nevertheless important that decision-makers who do so appreciate that past data has severe limitations in predicting the future. As Popper (1959) argued, we cannot prove that the sun will rise tomorrow merely because it has risen each day for centuries. However, we *can* say there is a *probability* that it will do so. It is also important for decision-makers to appreciate that the gathering of information to reduce *uncertainty* does not in itself reduce *risk*. The resulting information must be used to direct action to control risk. Risk management therefore involves a number of stages and we shall elaborate upon these in due course.

Second, there is an emphasis on events. This means that it is wrong to categorise risks as cost increases or delays because they are not events. These are the potential *impacts* or *consequences* of risk events. This simple but very important point is neglected in many risk management systems and texts, resulting in a focus on the consequences of risk events rather than the risk events themselves. Not surprisingly, organisations that categorise their risk profiles in this way are not very effective at managing their risks.

Third, there is an emphasis on the future. This is important because past events are not examples of risks, but are actual problems or crises that need to be resolved. Risk management is therefore a proactive process of looking forward and is fundamentally different from crisis management, which is reactive and backward looking. This distinction is also often confused in the construction industry where many managers think they are practising risk management when they are actually practising crisis management.

Finally, there is an emphasis on interests and objectives. Clearly, if a potential future event has no potential to adversely affect an organisation's objectives, then it is not a risk to that organisation. However, the same event may adversely affect another organisation's interests. This makes the concept of risk a very personal thing since an event that represents a risk to one organisation might not represent a risk to another. Indeed, an event, which is a risk to one organisation, might represent an opportunity (possibility of gain) to another, which can cause subsequent problems in securing teamwork to manage it. This is an endemic problem in the construction industry, which arises from the way in which construction contracts distribute risks. We shall return to the important distinction between risk and opportunity and to the problems of securing teamwork in managing them.

1.4.2   Risk and uncertainty

There has been much debate about the practical usefulness of the distinction between uncertainty and risk, which is often drawn in the risk management literature. It has been argued that the vast majority of risks facing a business cannot be accurately quantified because most, if not all, business decisions are made without the benefit of statistical data to enable an exact calculation of the impact of an event on an organisation's objectives. It is inevitable that in almost every business decision, there will be some informed *opinions, professional judgements* and *degrees of belief* about the event in question. Therefore, most expressions of risks will contain a degree of uncertainty in their calculations and it is possible to illustrate this distinction schematically in Figure 1.1, where risk and uncertainty are viewed at either ends of a continuum. In Figure 1.1, *risk* is at the assessable end of the continuum where there may be statistical data to produce an evaluation whereas *uncertainty* is at the less assessable end where evaluations will need to rely on informed opinion.

The distinction between risk and uncertainty is particularly relevant in relation to the management of health and safety. Here, despite increasingly stringent risk-related legislation, a lack of quantifiable data is the norm rather than the exception. This ensures that on most occasions managers are dealing with uncertainties rather than known risks.

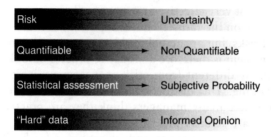

Figure 1.1 The risk-uncertainty continuum.

Despite the distinction drawn between risk and uncertainty, we prefer not to make the distinction in this book. Ultimately the distinction between risk and uncertainty is not one of substance, it is one of degree (of knowledge about the future event) and distinguishing between them serves academic rather than practical purposes. For this reason, in this book, we shall only use the term "risk" and we shall use it to mean:

> A potential future event which is uncertain in likelihood and consequence and if occurs could affect your company's ability to achieve its project objectives.

1.4.3   Probability and consequence

To understand the concept of risk it is important to understand what exactly is being measured. Risk is normally expressed in terms of *probabilities* and *consequences* (*impact on objectives*).

A probability is a number, between zero and one inclusive, which represents a judgement about the perceived relative likelihood of some event. For example, a company may say that there is a 1 per cent probability that there will be a delay on one of its projects. This means that they have past data, experience and opinions to predict that on 1 project out of every 100 projects there will be a delay.

To be a risk, a potential event must have a probability of between 0 and 1. A probability of 0 (0 per cent) implies the event is impossible and not a risk because it will certainly not happen. In contrast, a probability of 1 (100 per cent) implies the event is certain to happen. This makes it a real problem to be resolved rather than a risk.

An important issue to consider when interpreting probabilities is the method of measurement, which can result in quite different outcomes and therefore different decisions. For example, in times of economic boom and labour shortages, if one measured the risks of a fatality in the construction industry by counting the number of deaths per m^2 of floor area constructed, then the results might be very different to the number of deaths per person employed.

The first method would likely be preferred by the executive and indicate that the industry was getting safer but the second method would likely be preferred by the union delegate and indicate it is getting worse. Of course, any responsible and meaningful measure of risk should consider all perspectives but there is a strong natural tendency for people to perceive risk from their own perspective. It is a tendency that is difficult for people to resist or even recognise and we shall return to this issue in Chapter 3.

While important, probabilities alone only partially describe a risk and in isolation are of little value to decision-makers. For example, it is of little help to decision-makers who have to make a bid, to know that there is a 5 per cent probability that labour costs will increase. What allowance would one make in the bid to cover this eventuality? To make such an allowance, decision-makers would need to have information about the consequences of this event on costs. For example, a company may say that there is a 5 per cent probability that labour costs on a project will increase by $100,000. This means that they have data, experience or opinions to predict that on five projects out of every 100 projects labour costs will increase by $100,000. This risk can be expressed in terms of a single figure, which reflects its consequences for a project's budget. This can then be included as a contingency in that budget and is calculated by multiplying the probability of the event by the consequence if it occurred. In this case, the risk can be expressed as:

$$0.05 \times \$100,000 = \$5000$$
$$\text{Risk} = \text{Probability of event} \times \text{Magnitude of loss/gain}$$

It is important to consider both probability and consequences when assessing risk because although something may have a very low probability of occurring, extreme consequences can make it a very high risk. A good example of this is the radioactive leaks in the construction of a nuclear power station. The probability of this happening may be very low but the costs in human and monetary terms if it did occur would be enormous. So despite its very low probability, it is imperative that this risk be considered carefully.

History provides many examples to illustrate that in general we are not very good at identifying and managing low probability/high impact events. Indeed, even when seemingly obscure risks are identified and allowed for, it is all too easy to miss a crucial possibility. For example, in the design of the World Trade Towers, allowance was made for the possibility of an aircraft flying into the side of the building but that it would be fully laden with fuel and deliberately directed at a specific floor was completely unanticipated.

1.4.4   Probability and imminence

Another important distinction which is often confused when interpreting probabilities is the critical one between *probability* and *imminence*. It is important to realise that probability reflects the likelihood of an event occurring based on past experience or data, *not* the likely timing of that event. It is a common error to think that a one-in-a-million event cannot occur tomorrow. While a low probability event like this is less likely to occur than a high probability event, it is quite possible that it could occur first.

This is a critically important but often neglected distinction when companies are formulating a strategy to respond to risks. For example, when imminence is ignored, the natural tendency is to make high risks a higher priority than relatively lower risks. However, when imminence is considered, an imminent low level risk may justify higher priority than a non-imminent high level risk. While recognising the difficulties involved in measuring imminence, a minimum response should be to ensure that a considered plan is in place to deal with all identified risks, even if the plan is to do nothing. This is discussed more in chapter five.

1.4.5   Risk and opportunity

Risk can travel in two directions. Outcomes may be better (the upside of risk) or worse than expected (the downside of risk). Risk management should be as much about maximising opportunities for gain as it is about minimising the risks of failure. After all, companies make money by converting risks to opportunities. The distinction between risk and opportunity is therefore central to business success and is an important, distinctive and central theme of this book.

The fact that risk has an upside and downside is not new. For example, one of the most enduring classifications of risk in the risk management literature is the distinction between *pure* and *speculative* risks. Pure risks are those that offer only the prospect of loss whereas speculative risks are those that offer a chance of loss or gain. An example of a pure risk would be a client not following through with a project, the theft of property or the death of an employee. An example of a speculative risk is exchange rate fluctuations having an adverse or positive affect on an organisation's profits from overseas activities. While it could be argued that every risk has an upside and downside (a potential theft could bring about improvements in security and a potential accident could bring about improvements in working practices), the focus of risk management in the past has been almost exclusively on the management of pure risks. This is because the common usage of the word is almost exclusively negative. To most people, risk implies danger, which is why the term "hazard" is often used synonymously with risk, particularly in the area of health and safety. This has meant that in most instances, opportunities to improve performance beyond original expectations are often neglected or come as a by-product of the risk management process. However, if one is alert to opportunities at the outset of the risk management process, then many unexpected insights can be gained to improve performance. New opportunities may be identified in the course of managing risk and as a consequence of how risk is managed. The more rigorous and sound the risk management process the better the chance of identifying opportunities. Indeed, a focus on opportunities may be an important factor in determining effective risk control measures.

Another reason for the focus on the downside of risk, particularly in the construction industry, has been the penal nature of many contracts, which traditionally place far more emphasis on punishment for underperformance than on rewards for over performance. Recently, many industries, including construction, have seen the introduction of legislation underpinned by the prospect of severe fines or even imprisonment for breaches of compliance. There is therefore much more of an incentive to look for problems and prevent them than to seek opportunities and take advantage of them. Consequently, most companies focus on the downside of risk, despite the rhetoric of their system documentation which may refer to the need to take opportunities. One reliable indicator of this negative mindset is the vocabulary used in almost every company's risk management systems. For example, the use of the term "catastrophic" is almost universal in company risk matrices to describe the most extreme consequences of a risk eventuating. Yet the idea of a catastrophic opportunity makes no sense. The result has inevitably been a focus on minimising hazards rather than maximising opportunities, mediocre rather than optimal performance and a culture of mitigation rather than innovation. The negative impact that this dominant culture of mediocrity has had on the design, construction and management of buildings is likely to be significant and should not be underestimated.

There has been some vigorous debate about whether the concept of risk should encompass both opportunities and threats or whether risk should remain an exclusively negative concept, separate to that of opportunities (Hillson 2002, Loosemore and Lam 2004). The problem for an organisation in making risk an umbrella term, which encompasses both threat and opportunity, is that people focus on the former and ignore the latter. This is the reality for the majority of organisations in many industries and the inevitable result is lost opportunities and lower performance (Bowden *et al.* 2001, Chapman and Ward 2002, Loosemore and Lam 2004). Conversely, the problem with separating the concepts of risk and opportunity and treating them differently is the inefficiency which could result from the extra administrative burden that this could involve for already busy managers. Once again, the result could be lost opportunities. As Hillson (2002) argues, the best solution which allows both opportunities and threats to be managed proactively is have them imbedded in the same system and managed by the same processes with some modifications to reflect the different ways in which threats and opportunities need managing. This makes sense because risk and opportunity are synonymous – the opposite sides of the same coin. Every risk has an opportunity and every opportunity has a risk. To treat a decision as an opportunity in isolation from its risks or vice versa would be nonsensical and potentially dangerous. Nevertheless, this is what normally happens. Unfortunately, the decision of how to define risk and opportunity is one that is glossed over and taken for granted by most organisations but it is clear that it is an important distinction which must be considered carefully. As Bowden *et al.* (2001) point out, businesses that want to be sustainable in the twenty first century need to adopt the philosophy that risk management is a process of continuous improvement directed towards the identification and effective management of opportunities as well as potential threats. For this reason, we shall distinguish between the terms "risk" and "opportunity" for the rest of this book. *Risk* will be used to refer to potential events which could adversely affect a company's objectives and *opportunity* will refer to potential events which could beneficially affect a company's objectives.

1.4.6   Objectives

"Risk" and "opportunity" are simply words that we use to express the impact of an uncertain future event on specific individual, group or organisational objectives. For example, if one objective is to complete a contract on time, then a company's risk profile includes all of those uncertain events which can affect that objective negatively. If that objective changes, then other uncertain future events become risks and need to be managed.

The concepts of risk and opportunity become complicated when one considers the wide array of objectives which drive modern business organisations and how these change over time. It becomes even more complex when one considers the many stakeholders with an interest in today's projects and the widely varying objectives they may wish to achieve. This means that the risks and opportunities associated with a project are likely to be perceived in many different ways. For example, the impact of a dam project on the environment is likely to be perceived very differently by the engineer who designed it, the local residents downstream whose lives depend on it and those who live upstream who may be facing compulsory purchase and relocation. While the local people's perceptions might be at variance with the scientific evidence presented by engineers, to the stakeholder they appear to be reality and this is all that matters to them. Unfortunately, there is a tendency for managers to focus on measurable risks and scientific evidence, ignoring stakeholder perceptions. This is both an ethical and commercial problem and it is an issue to which we shall return.

It is clear from the above discussion that risk and opportunity are simply measures of *personal* exposure to uncertainty. This means that their identification and assessment is an inherently subjective process. Consequently, people's responses to risk are less likely to be based on actual risks, than on their perceptions of them. These perceptions are in turn determined by people's personal experiences, beliefs, values, attitudes and social networks (Bowden *et al.* 2001, Edkins and Millan 2003). For example, an individual is more likely to oppose such a project if they have been exposed to infromation or an experience which provides evidence of environmental damage or are imbedded in a social network which reinforces these beliefs. Therefore, even if a project appears worthy of acceptance, it may not necessarily be accepted by stakeholders. This is a problem which has been experienced by many construction and engineering companies. Later in this book we will return to the important subject of public risk perceptions and how a company can manage them effectively.

1.4.7   Responsibilities – Voluntary and involuntary risks

On a project, the potential risks and opportunities to which an organisation is exposed are largely determined by the responsibilities it has accepted under the array of legal contracts signed with its business partners. For example, if a contractor contracts with a client to deliver a building within a certain budget, then it accepts the responsibility for managing those events (risks) which can potentially cause that budget to be exceeded. Of course, on the upside, the contractor also accepts responsibility for the potential events (opportunities) which can reduce costs under budget. The important point is that these events are not risks or opportunities to the other companies involved in the project, which have not signed this particular contract. These companies will be exposed to risks and opportunities of their own, according to their contracts.

Overall, a project manager responsible for an entire project would hope that the variety of contracts used on a project would collectively assign responsibilities to manage *all* project risks. However, in reality, due to large numbers of parties involved in projects, *ad hoc* amendments to contracts, misunderstandings of contracts and poorly drafted contracts, this is seldom the case. Hence, some risks will inevitably go unmanaged.

To avoid this, it is useful to distinguish between the typical contractual risks and opportunities which different construction project participants accept responsibility for managing (Edwards 1995). For example, a promoter (financier or client) who has to justify investment decisions to shareholders would normally be concerned that a reasonable rate of return is obtained from the project. They may also be concerned with the effective functioning of the building for its intended use. Typical risks taken by promoters which can impact on these objectives might include broad management responsibilities for consultants, changes in requirements, project finance, the timeliness of payments, approvals, public protest, war, national industrial action, terrorism, changes in regulations and legislation, unknown ground conditions, use of existing services, access to site and natural disaster etc. In contrast, traditional contractors and sub-contractors will be exposed to risks affecting shorter-term cash flows, programmes and quality issues and will be more concerned than the promoter with the technical risks associated with specific packages of work on site. Typical risks accepted by contractors might include fire, theft, poor weather, labour and materials availability, labour and material cost increases, sub-contractor performance, poor craftsmanship, poor programming, poor pricing in tenders and strikes or labour disputes on site. Other parties, such as professional consultants will view project risks from a different perspective again. Their risks would typically include the accuracy of fee estimates, good management, prompt and full fee payments from clients, possible claims of negligence and insurance premiums. Finally, third parties such as insurers will only be concerned with the risks they have covered. Their primary concern will be that they have accurately estimated those risks, imposed the correct insurance limits and conditions and charged the correct premiums.

In addition to the *voluntary* risks which organisations sign up to in their contracts of service, all companies are also exposed to *involuntary* risks which are associated with their responsibilities implied by common law, legislation and codes of conduct issued by regulatory institutions and moral codes of conduct in society at large. Involuntary risks are also passed down the procurement chain by those who have not managed their own risks effectively or by those who pass on risks subversively without consent or knowledge. Unfortunately, this is common in the construction industry and many companies either fail to manage these risks or are prevented from doing so by established practices. For example, many contractors, sub-contractors and facilities managers are excluded from the design process. The inevitable result, when organisations discover they are exposed to these involuntary risks, is delay, financial loss and conflict (Loosemore 2000).

1.5   Risk management maturity

Having explained the common terminology underpinning risk management, we turn now to discuss the idea of risk management maturity. This is reflected by the sophistication of an organisation's understanding of its risk portfolio, its knowledge of how to mitigate those risks and of the extent of its internal business continuity systems needed to cope with and recover from risk events. Establishing the maturity of risk management in an organisation is a useful starting point when embarking on a review of current risk management practices, systems and culture. To this end, the work of Mitroff and Pearson (1993), Pearson *et al.* (1997) and Pearson and Clair (1998) is of particular value. They found that organisations, which are immature in their approach to risk management, have certain characteristics in common.

1.5.1   Characteristics of risk-immature organisations

Stereotypically, risk-immature organisations tend to be sceptical about risk management and are often characterised by a culture of success and managerial invincibility. They also tend to have task-orientated cultures which consistently stress the importance of profits over people and other corporate goals. The result is highly geared, overly lean organisations, which have no spare capacity to deal with the unexpected. These organisations exist "on the edge" and nurture a mindset that company size and past successes provide protection from future risks, that problems happen to others, that good management and hard work prevents problems and that desirable business ends justify the taking of high-risk business means. People in risk-immature organisations often believe that that risk management is someone else's responsibility and that they have the power to offload risks onto other parties, thereby insulating themselves from the uncertainty of their environment. For these types of organisations, risk management is considered a sign of weakness because problems are seen as a sign of managerial failure. Furthermore, there is a reluctance to re-examine existing organisational practices in the aftermath of a problem and rather than learning lessons for the future. Instead, the priority is to maintain the organisation's public image and to ensure that internal operations remain intact. In essence, the risk management systems of immature represent little more than a managerial façade to impress external stakeholders and reassure managers that something is in place to deal with the unexpected, even though they know that they have minimal impact upon day-to-day organisational practices and attitudes.

Pearson *et al.* (1997) found that these characteristics defined the way that many organisations across a range of industries operate. This is supported by recent cross-sector research in Australia which found that while 83 per cent of organisations with over 100 employees had a business continuity plan, less than half had ever tested or reviewed them to maintain their relevance (Timson 2003). Similarly, in the UK, only 47 per cent of all organisations surveyed by the Chartered Management Institute and Business Continuity Institute in 2004 had Business Continuity Plans in Place. Furthermore, only 57 per cent of this group tested their plans annually and only four kinds of disruption (loss of IT, Telecommunications, site and fire) were typically covered. Levels of knowledge about risk management are also a concern. For example, in a recent survey of company directors across a range of Australian industries Stewart (2004) found risk management training had been undertaken by only 29 per cent of directors, 49 per cent of CEOs and 47 per cent of senior managers. As one would expect, standards and methods of risk management and the types of risks identified in different sectors varied considerably. For example, the major risks identified in the public sector were environmental, legal and security. They also tended to be identified and analysed by a risk management department and reported to senior management by an audit committee against specific targets in an annual risk management plan. This created the potential problem that "softer" risks not normally addressed by audits went undetected. Such risks might include actions by activists, loss of specialised staff, product liability etc. Stewart (2004) found that the public sector organisations that performed best were those that had created a risk management committee at senior level with responsibility for ensuring "significant" risks were declared to the board for consideration. Without this mechanism, many senior managers were ignorant of the risks facing their organisation. In contrast to the public sector, the main risks identified by the private sector were occupational health and safety, public liability, competition and financial risks. While the CEO tends to have primary responsibility for risk mitigation, all managers are expected to identify and evaluate risks, particularly in the consulting sector. The principal method of reporting to the board was the CEO report and private companies were less likely than public companies to have a formal risk management policy and independent risk reporting channels to the board.

This portrait of widespread risk management immaturity appears to be reflected in the construction industry. For example, recent research into the risk management practices of the UK's top 75 construction companies revealed that 38 per cent of companies were not confident in their risk management strategies, 21 per cent did not have a director with responsibility for risk, only 41 per cent reviewed risks on a monthly basis and 21 per cent did not have a risk-based approach to tendering (Smee 2002). Fifty-seven per cent of these companies regularly declined tenders on the grounds that they were too risky. Other research has indicated that even on highly complex PPP/PFI projects, risk management practices are highly variable, intuitive, subjective and unsophisticated (Akintoye *et al*. 2001, Cottle 2003). Finally, in a study of the crisis preparedness of major Australian construction companies, Loosemore and Teo (2002) found that corporate philosophies do not support crisis management and that the limited planning that does occur is undertaken in an insular, informal and haphazard manner, supported by little strategic guidance and resources. If this is the standard of risk management in the construction industry's biggest firms, imagine what the standard of risk management is likely to be in the 90 per cent of firms that employ fewer than 10 people? Collectively, this research indicates that while many construction firms might have risk management procedures and systems in place, it is questionable whether they can deliver real benefits in today's volatile conditions.

1.5.2   Risk-mature organisations

In contrast to risk-immature organisations, risk-mature organisations typically have a culture of openness, awareness and sensitivity to organisational risks and of their social and financial responsibilities to stakeholders, the general public and the wider environment (Ginn 1989, Lerbinger 1997, Pearson *et al*. 1997). In such organisations, proactive risk management is systematically incorporated into strategic planning processes and championed by senior executives so that it is an integral and instinctive aspect of organisational life at all levels (More 1995). This is particularly important in relation to health and safety where senior management commitment to good performance is fundamental to others in the organisation carrying out their health and safety responsibilities and may be required to be demonstrated in law. Senior executives also support risk management by providing sufficient resources and clear statements of fundamentally held, core beliefs and attitudes relating to organisational priorities. In addition, large organisations tend to have a permanent risk management team, charged with the responsibility to create a comprehensive risk management plan and to continuously communicate, coordinate and review risk management efforts. There is also flexibility and willingness to "let-go" of formal, standardised systems and procedures which serve them well in "normal" times but which can become restrictive and counter-productive during a risk event (Sagan 1991). This requires a capacity to communicate effectively with external and internal stakeholders at a time when formal information systems can become stretched and overloaded with information.

During a risk event, effective communication is essential but often difficult and it seems that companies with a track-record of effective communication as an intrinsic part of their day-to-day life are most likely to survive (Mindszenthy *et al.* 1988, Aspery and Woodhouse 1992, Sikich 1993). Effective communication systems are particularly important during a major crisis which usually involves non-routine dealings with external stakeholders such as emergency services, the public, the media and existing and potential customers. Being able to deal with the media is critically important because they play a crucial role in shaping the public's image of events and perceptions of blame for any undesirable consequences. Poor communications can result in distortions of the truth, unjustified mistrust, suspicion and irrevocable damage to public and customer relations.

Risk-mature organisations also understand the interdependence of risks and encourage collective responsibility for the management of those risks between everyone involved in their supply chains. This is particularly important in the construction industry where supply chains are long, fragmented and unwieldy. It requires a willingness to share risks appropriately along a chain in a way which ensures that those who bear risks have the knowledge and resources to control them. Unfortunately, the dominant culture in construction is risk transfer rather than risk sharing. In construction the normal practice is for each member of the supply chain to pass on risks using back-to-back contracts to the point of least resistance which is normally the organisation least able to cope with the risk. Too often the result is tension, resentment, conflict and a failure to take responsibility for problems when and where they arise. This forces people into a reactive mode of risk management, having to respond to preventable risks which have inevitably grown in proportion and become more difficult to manage. For example, Bea (1994) found that on some projects in the US, up to 60 per cent of the problems that arise during construction have their origins in the design phase. In Australia, the National occupational Health and Safety Commission attributed 46.5 per cent of fatalities between 1997 and 2002 to definite or probable poor design and 16.3 per cent to possible poor design (NOHSC 2002). Changing this culture in construction has proved difficult for many decades because of deeply imbedded and hierarchical professional divisions which are reinforced and cemented into place by complex, legalistic contracts and fragmented procurement systems.)

With the increasing popularity of Public Private Partnership (PPP) and Private Finance Initiative (PFI) projects there is suddenly a new incentive to change the construction industry's traditional reactive and confrontational practices and drive major efficiencies in the management of risks along supply chains. PPP and PFI projects ensure that contractors and consultants can no longer walk away from the projects they build. They also pass new risks to the private sector for the entire life cycle of a building and by doing so encourage a life cycle approach to the management of project risks which in turn requires hitherto competitive firms, business units and functions to work together. Indeed, by making project teams the customers of their own services and products, PPP and PFI projects are likely to lead to a complete reversal of power in the supply chain where those at the operational end drive the production end. In a PPP project, clients make no payments during the development period and only start making payments when there is "commercial acceptance" that the building is completed enough to enable service delivery to the specified standards. So a consortium taking the risk of operating a facility at a certain level of service for 25 years will depend on service income generated during the operating period, rather than on the production income generated during the construction phase as in traditional contracts. This means that facilities managers, who in the past have been largely excluded from the building process, will now need to be intimately involved in the relatively short period of design and construction activity that will determine the consortium's annuity income for much longer 25-year operating period. For example, in one recent 25-year PPP project to construct the new hospital in Sydney, Multiplex Facilities Management placed two senior facilities managers on site for the two-year design and construction period to provide advice to designers and contractors on the operational implications of their decisions. This gave confidence to the client and the consortium that the building would be fully operational from practical completion and allowed Multiplex Facilities Management to effectively manage the risks and opportunities associated with the 107 key performance indicators (KPIs) on the project. It also ensured that the construction and design teams were aware of these operational KPIs and could take them into account when making their own decisions. This idea has been labelled "common horizons" to reflect the ultimate objective of encouraging decision-makers in different stages of the procurement process people to appreciate the interrelationship between their KPIs. The result is a better solution for every stakeholder involved in a project.

1.6   Diagnosing risk management maturity

A useful diagnostic tool for measuring risk management maturity, which has been specifically designed for projects, is the Project Management Institute's Risk Management Maturity Level Audit Tool (PMI 2002). This tool categorises an organisation's risk management systems and practices into four levels of maturity. Level 1 is the lowest level of maturity where an organisation's risk management practices and systems are largely *ad hoc*, unstructured and reactive. In such an organisation, there are no dedicated resources for dealing with risk, no formal risk management processes, no risk awareness and no attempt to learn from past projects or prepare for future projects.

At Level 2, there is still no structured approach but there is some experimentation with risk management by a small number of people on selected projects with little consistency. Risk management is viewed as an additional overhead and, although upper management might encourage the use of risk management, it is not enforced, there is no formal training and there are no generic processes in place. At Level 3, there are dedicated resources for risk management which are integrated into organisational processes on most projects, through a formalised and generic risk management process with specific processes and tools which are also integrated into quality management processes. Furthermore, there is a clear policy for risk management and risk data is collected and analysed by an in-house core of expertise with little need for external assistance.

Level 4 is difficult to achieve and needs significant investments of time and resources. It is characterised by a proactive culture of risk management which is inextricably integrated into every project, organisational function and supply chain. State-of-the-art techniques are used to identify and analyse risks and there is top-down commitment to risk management. Furthermore, innovations that exploit the best risk management practices are continually developed, identified and transferred throughout the organisation. Customers engage actively in the risk management process and employees "think" risk management and follow generic and well-established processes on an automatic basis. There is also active use of information to improve organisational processes, learn and disseminate lessons and gain competitive advantage. Finally, all staff are regularly trained in risk management and are capable of using basic risk management techniques.

While valuable, the Project Management Institute's Risk Management Maturity Level Audit Tool is quite narrow in its description of what characterises each level of maturity. It also needs refining to suit the peculiarities of different industries such as construction. Indeed, when combined with other research in this area, it is possible to produce a more robust model which can better indicate an organisation's level of risk management maturity. This is illustrated in Appendix A which lists the typical attributes of an organisation at each level of maturity under the headings of; *awareness, culture, processes, skills/experience, image, application, confidence* and *resources.* Appendix A is based on an integration of work by Mitroff and Pearson's (1993) and Loosemore (2000) and indicates the types of questions which should be asked when assessing risk management maturity. It can also be adapted by organisations to suit the peculiarities of their own business.

Experience in using this audit tool indicates that most companies will be operating at different levels of maturity for different types of risk. For example, an organisation may find that it is operating at Level 4 with its financial risk management practices but at Level 1 with its environmental risks. An organisation can also operate at different levels of maturity in the different maturity categories. For example, an organisation may have a Level 1 culture but Level 3 processes. In other words, a company may have developed a sophisticated system but not imbedded it within its organisational behaviour and practices. The challenge for any organisation is to achieve a consistent level of maturity across all categories and across its entire risk portfolio. Few organisations will achieve Level 4 in all aspects of its business but a useful way to monitor and illustrate progress internally and externally is to plot one's position in each dimension using a simple spider diagram as illustrated in Figure 1.2. In this typical example, the organisation in question has conducted two audits at different times in its development – as represented by the inside and outside lines. If the outside

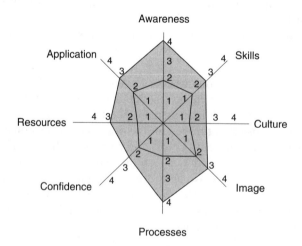

Figure 1.2 Spider diagram of risk management maturity for all risks.

line is the latest audit result, this company is clearly making progress in building awareness and processes but less progress in developing skills and a risk management culture. In using this audit tool, companies can draw a spider diagram like this for its entire risk portfolio or, if it wants to be more precise, it can draw a spider diagram for each type of risk (time, cost, environment, health and safety, quality etc).

1.7 The increasing importance of community consultation and managing risk perceptions

Risk has always been part of our human existence and has been traditionally explained and dealt with at a community level through folk law, taboos, rituals, superstition, religion, myth and magic. For example, the Islamic term "by the will of Allah" illustrates the way in which many religions deal with the threat of disaster or death by attributing such events to divine intervention. Although the development of science has helped us realise that we can control our fate, it is only recently, in the post-war years, that the scientific study of risk emerged. This began when probabilistic tools of safety analysis were designed for the first space exploration programmes and nuclear power plants. The subsequent emergence of risk management as a profession in its own right has been characterised by the translation of responsibility for risk from the community to a professional bureaucracy such as a government regulatory body, with decisions relating to societal safety being made centrally at an institutional level (Barnes 2002). This trend not only emerged in response to technological advances which brought many new risks which necessitated independent control, but was also part of broad-based societal change, linked to the rise of the centralised nation state as the dominant political institution and the subsequent success of centralised public health initiatives. While inevitable to some extent, the problem with this trend is that risk experts in regulatory bodies and academia have developed their own body of knowledge, linguistic conventions and practices which have isolated them from the lay public who reside in the community.

The scientific, technical and centralised approach to risk management which often appears to devalue and ignore public perceptions of risk has led to a perception amongst the public that decisions affecting their lives are out of their control. The resultant lack of trust and confidence in business and regulatory institutions has been fuelled by numerous instances of scientists, companies and regulatory bodies covering up key risk information, being wrong in their assessments of risk or not communicating them effectively. Good examples are the initial attempt at covering up the BSE crisis in the UK and the SARS and bird flu epidemics in Asia and more recently the James Hardy debacle. This type of approach to risk communication is no longer acceptable or sustainable. While once it would have been thought that the public were exhibiting irrational behaviour or had a deficient understanding of the "facts", the empowerment of the masses through better education and the spread of democracy ensures that regulatory bodies and industrial organisations can no longer afford to ignore their perceptions and must imbed these in their decisions. This does not mean that managers should abandon technical risk analysis. After all, if reliable data is available on risks, then it would be wrong to ignore it. The point is, that risk analysis should not be confined to objective outcomes and measures but should take account of normal people's perceptions of risk. To achieve this, effective and meaningful community consultation must be recognised as a crucial part of effective risk management processes. Furthermore, managers should appreciate that scientific and technical assessments of risks by experts are relevant only to the extent that they are integrated into individual perceptions. People respond to decisions according to their perception of risk and not necessarily according to objective measures. This is well illustrated by Wilson and Crouch (2001) who point out that while statistics prove that Britain is a safer and more prosperous place to live than at any time during its history, the British public have a gross misconception that health risk levels in their society have never been higher. This tells us that in addition to reflecting an absence of harm safety can be described as the social judgement that the probability of a risk is appropriately low or that the consequences of a risk is acceptable to the limits set by society. It also implies that the establishment of safer communities and workforces is not dependent entirely on technical advances but on understanding what safety means to those communities themselves. Therefore, the emphasis of risk management strategies should not only be on technical solutions but also be on developing a better understanding of stakeholder values and of how they perceive the risks created by industrial activities. The management of health and safety risks should ideally be supported by an industry culture that reflects society's values, which places a premium on good health and safety performance and good industrial relations.

It would seem that the key to better managing the publics' perception of risk is fundamentally one of communication, consultation and involvement in decision-making. It is also about identifying and understanding the stakeholders in a business and what is likely to influence their perceptions of risk. The key point is that objective, actuarial and technical measures of risk, although easier to operationalise, have little meaning if they are separated from the social and behavioural context in which risk is experienced by the public. This is not a theoretical idea or ideal, as is demonstrated by following quotation from the Executive Director of the Australian Victorian state government's Treasury and Finance Department's PPP "Gateway" Initiative:

> I want to emphasise at the outset the importance of positioning, of having an overarching theme or brand for the project. It is not enough in the

communications field to simply tell people what is happening; that it is essential, but that it is only half the job. To help to ensure that the project is to proceed smoothly to completion and fulfil the government's public relations needs, the community at large must feel positively about the outcomes and develop an expectation about the benefits that will flow. To achieve this, it is essential to develop a public vision for the project and set of key messages to be incorporated into all significant public information activities.

(Sharpe 2004: 8)

Indeed, this need for meaningful community consultation about risk has echoed through the risk management literature for over a decade; and over this period we have begun to develop an understanding of how lay people think about risk and what shapes their perceptions. First, we have learnt to realise that anybody is entitled to act in a way which may appear to others as inconsistent, irrational, illogical and unreasonable within the law. Second, we now appreciate that the language used in communication with stakeholders needs to assist communication and not be a medium of dominance disguised by indecipherable jargon. It is critical not to talk down to stakeholders or to appear to hide anything from them. Third, communication needs to be based on a mutual sense of respect for the positions and perspectives of all parties. It is important for any organisation to know its stakeholders, what they want to know, what they want to achieve, what their preconceptions are, whether they are really interested in your message or are simply trying to tell you something. If it is the latter, then listening is probably more important than talking. Fourth, we now know that risk management should be a multi-way process that is designed to make information about risk accessible to everyone and to promote mutual understanding, if not consensus. Presenting a *fait accompli* to staff in an organisation or indeed, to the public, is almost certain to result in objections. It is therefore essential that consultation processes that facilitate two-way communication are employed. Finally, it is important to ensure that risk assessments are not presented as a single option for action but as a presentation of facts or judgements to enable the audience to decide for themselves from a range of options. As Fischhoff (1995) says, the key to the effective management of stakeholder perceptions is to get the numbers right, tell them the numbers, explain what the numbers mean, show them that they have accepted similar risks in the past if this is the case, show them that it is a good deal for them if this is the case, treat them with respect and make them partners. Following these general principles does not guarantee the elimination of conflict between organisations and their external stakeholders, but it does offer an opportunity for them to become part of the solution and to gain a better insight into the constraints under which decisions are made, even if they do not necessarily agree with the eventual outcome. For example, the rights of trade unions to be involved in decisions that affect their members' welfare is for some a contentious issue. However, there is evidence to show that when trade unions are effectively engaged, they play a critically important role in establishing and maintaining safe *and* effective working conditions for employees (Loosemore and McGeorge 2002, Loosemore *et al.* 2004). Unfortunately, politics and imbalances of power between stakeholders often prevent appropriate involvement in decision-making processes. When this is the case, a far greater onus is placed on the risk management team to provide a transparently fair process to reflect stakeholder views.

1.8 Contemporary debates in risk management

The diverse and relatively immature field of risk management has been described as an archipelago of competing doctrines (Hood and Jones 1996). The seven most prominent contentions in risk management are illustrated in Table 1.1, which defines a number of extreme positions that can be taken in developing a risk management approach. Most managers exist between these extremes and the position one takes depends upon one's beliefs which are in turn shaped by one's culture, personality, education and experiences.

In essence, the two columns in Table 1.1 represent the two main schools of thought discernible in current risk management literature. These are called the *homeostatic* and *callibrationist* viewpoints. The jargon is irrelevant but the differences between these two camps are very important. Essentially, the homeostatic view (doctrine column) represents current orthodoxy in risk management practice. It emphasises prevention rather than cure, anticipation, quantification and the specification of precise outputs. It assumes that all organisations are capable of setting clear and measurable goals and translating them into standard operating procedures. In contrast, those that hold the callibrationist view (counter-doctrine column) argue that reliable forecasting is not possible in many business areas. Furthermore, the variety of stakeholders in an organization makes it difficult to establish common goals that can be translated into a single set of procedures. The callibrationist view also argues that in an increasingly uncertain world, organisational resilience (an ability to respond to problems and opportunities after they have arisen) is becoming more efficient than an emphasis on prevention. Callibrationists believe that while prevention is better than cure, it is impossible to create a crisis-free environment. An overemphasis on prevention is therefore futile, inefficient and even dangerous. Crises are becoming an inevitable and indeed positive aspect of organisational life, if they can be managed effectively.

1.8.1 Rationality and risk quantification

Underlying the two broad positions identified in Table 1.1 is the greatest and most enduring controversy in risk management – the issue of *rationality*

Table 1.1 The seven most prominent contentions in risk management

Doctrine	Justification	Counter-doctrine	Justification
Anticipation	Prevention is better than cure. It is possible to apply causal knowledge of previous system failures to better guide future risk management decisions.	Reaction	Complex system failures are not predictable and it is impossible to create a crisis-free organisation. Too much emphasis upon prevention makes things worse by creating a sense of invincibility.
Blamism	Targeted blame for unexpected problems gives strong incentives for taking care on the part of key decision-makers.	Absolution	A "no-blame" approach to unexpected problems avoids conflict, recrimination and the distortion of information, thereby helping risk management activities.
Quantification	Quantification promotes understanding. A rational risk management system must rest upon systematic attempts to measure risk.	Qualitativism	Proper weight needs to be given to the inherently un-quantifiable factors in risk management.
Reliable knowledge	We understand well, how organisational factors influence risk meaning that orthodox engineering (cause and effect) approaches to risk management are reliable in organisations.	Unreliable knowledge	There are major limitations in the current understanding of the factors influencing risk and how it is handled in human systems.
Independence	Safety must be explicitly traded-off against other goals.	Interdependence	Safety and other goals go hand in hand under good management.
Confinement	Discussion is most effective when confined to experts.	Consultation	Broader discussions better test assumptions and avoid errors.
Structures and products	Risk is best minimised by changing physical structures. The emphasis of risk management should be on providing precise goal statements and control systems relating to product specifications.	People and processes	Risk is better coped with by changing the behaviour of individuals and the emphasis of risk management should be on specifying processes rather than physical standards.

Source: Adopted from Hood and Jones (1996).

and *risk quantification*. Probably more than any other development, it is the quantification of risk and the rise of *scientific rationalism* that have distinguished modern society from the rest of history. Herein lies the traditional actuarial world of risk management with strong traditions in the mathematics of probability. However, Bernstein (1996) has questioned the extent to which we have gained from the superstition to supercomputer transition, where probability analysis has supplanted hunches and intuition in many areas of human endeavour. While it is indisputable that the development of scientific techniques has enhanced our quality of life by allowing people to take more calculated risks than ever before, there is also a negative side. For example, Richardson (1996) argued that the mathematically driven devices of modern business can often relegate people to be a nuisance rather than a valuable resource. Furthermore, the idea that predictive business models can be constructed with precise probabilities is overly simplistic. In contrast, the real world is an *open system* which is not reducible to a neat set of relationships with an underlying order that is capable of being specified. According to Berry (2000), the real world is a trap for logisticians and the mathematical regularity they see is an illusion which hides the wildness laying in wait. There is no better illustration of this underlying wildness than the swings of the stock-exchange. Consider, for example, the months preceding the "tech-wreck" of early 2000 where unpredictable and vicious price swings in stocks followed many months of illogical investment in high-tech companies, which became grossly over-valued in comparison to their asset values and profit potential. It is impossible in the long, medium or even short term for computers or mathematical models to predict such irrational behaviour with accuracy and it is foolish to believe that we can attribute reliable figures and numbers to it.

1.8.2   The risks of rationality and risk quantification

The tendency for many managers to place unquestioning faith in quantitative risk management techniques encourages them to take risks, which they would otherwise not take, under the illusion that they understand them better. For example, it has been shown that the development of safer technology for cars can encourage some drivers to behave more aggressively. The result is that the number of accidents may rise even though the severity decreases. Coincidentally, this is exactly what happened in response to the introduction of the UK's CDM regulations, arguably the most significant risk-related legislation to hit the UK construction industry (CDM 1994). In this way, the science of risk management is capable of creating new invisible and therefore highly dangerous risks of its own. Adams (1995) explains this behaviour by arguing that people have a natural propensity to take risks and that increased perceptions of safety can lead to compensating behaviours to bring risks back to norm levels. Extending this logic to the construction industry where repeated attempts to improve safety on sites have met with varying degrees of success suggests that it might have developed a culture with a high propensity to risk, thereby causing people to adjust their behaviours in response to any risk-reduction initiatives. The challenge then becomes that of changing attitudes towards safety as well as putting technical devices in place to make the workplace safer. Of course, a balanced approach is appropriate. However, it is clear that risk management cannot be seen simply as a technocratic calculus divorced from its cultural context but as a process influenced by people's social relationships, preferences and beliefs. The operating climate that best supports this approach is one based on management practices that promote issues such as mutual trust, respect for the rights of individuals and open sharing of information. If management practices do not reflect some form of balance between the interests of an employer and its employees, effective consultation can be difficult to achieve. Since this is not apparent in some organisations, it becomes difficult to argue that there is not a legitimate role for other parties such as governments and trade unions in helping to establish and maintain the balance of interests necessary for effective consultation.

1.9   The risk and opportunity management process

In practical terms, risk management could be described as the process of proactively working with stakeholders to minimise the risks and maximise the opportunities associated with project decisions. The aim is not to avoid risk but to take calculated risks, make more informed decisions, avoid unpleasant surprises, identify opportunities and encourage people to think more carefully about the consequences of their decisions. Ultimately, the aim of risk management is to ensure that project objectives are achieved and ideally exceeded and that there is a consistency and reliability of service which leads to greater customer confidence. This involves developing rigorous processes to encourage managers to think through their decisions by identifying in advance, potential hazards and opportunities, assessing them and putting plans in place to mitigate and maximise them respectively. Such processes have to be developed by the people who work in an organisation. Purchasing off-the-shelf software, stuffing it with data and pressing a few buttons cannot effectively manage risks. The best risk management systems have been developed in close consultation with stakeholders and integrate seamlessly with existing business systems, activities and cultures to ensure that risk management is not seen as an extra burden but as something which is an integral aspect of day-to-day decision-making, implementation, reporting and auditing processes.

While many managers routinely practice risk management when making decisions, approaches often vary, resulting in inconsistent and unsystematic processes within and between companies. Furthermore, managers naturally tend to focus on those narrow areas of risk with which they are familiar or which seem directly relevant to core business activities. Consequently, and dangerously, many managers have a poor understanding of the risks associated with the broader operational aspects of their business, even though these events may present the possibility of considerable exposure. A systemised application of management policies, processes and procedures to the principles of risk management will help to avoid these problems by ensuring that risk management is practised in a structured, appropriate and consistent manner across all organisational levels and functions and during all stages of a project's life cycle. It will also ensure that decisions which are non-routine, complex and challenging receive particular attention and that routine day-to-day decisions receive a less intensive approach. This will help achieve better project outcomes and more efficient decision-making processes.

Broadly speaking, the three stages of the risk management process are:

1   Risk identification – Identifying risks and opportunities to decision objectives.
2   Risk analysis/assessment – Measuring the magnitude of the risks and opportunities identified.
3   Risk control – Putting controls in place to maximise potential opportunities and minimise potential risks. Monitoring those controls to ensure they have their desired effect, reviewing them and responding if they do not and learning lessons for future decisions.

Each of these stages will be discussed in detail in the following chapters. However, before progressing, it is worth emphasising the importance of effectively managing the risk management process. Organisational deadlines for reports such as business plans, corporate reporting and regulatory requirements, board meetings and internal and external project reviews can often set the pace and timing of the risk management process. While these are important deadlines for which risk reporting may be central, it is crucial to see risk management as a continuous and value-adding process aimed at supporting and enhancing ongoing internal business operations not as a periodic and cynical process of keeping external constituents happy. Furthermore, adequate time and resources must be allocated to the process, which can take from a few hours (for simple decisions) to several months (for complete business risk assessments) and involve considerable effort from many contributors. One of the largest investments will be in the provision of training to ensure that staff and external stakeholders understand the importance, relevance and value of effective risk management, that they have the skills and knowledge to effectively practice it and that they understand the systems and procedures which ensure that this is done consistently and systematically.

1.10   Conclusion

In this chapter we have argued that risk management, although a very valuable process for dealing with this increased level of risk, is not a precise science or, indeed, a particularly well-developed art form. We have also highlighted the danger in seeing risk management as a complex mathematical process, underpinned by precise scientific tools which can provide accurate assessments of risks and flawless strategies for dealing with them. Effective risk management is most fundamentally a human process of systematic, rigorous and creative thinking underpinned by some simple tools and techniques. Furthermore, the magnitude of risk for an organisation is a function of the quality of its social relationships and processes, as well as the technologies it uses in its activities. Organisations which recognise the limitations of numbers in risk decision-making and become more attuned to the political, social, emotional and ethical aspects of risk management are far more likely to understand the full diversity of risks facing them.

In his seminal and now prophetic text, Beck (1992) argued that we live in a "risk society" where science and technology will create additional hazards to those which naturally occur and of which we are only becoming vaguely aware. The global media coverage and widespread knowledge of, and public engagement with the terrorist incident of September 11th 2001 has caused a sharpened realisation that it is not just science and technology which is creating new risks for society. It is also globalisation, limited resources and inequalities in the way that wealth and power is distributed in the world. In this new era of risk sensitivity, risk management has moved from a traditional compliance orientation to become a critically important and essential tool to ensure organisational survival, corporate responsibility and to secure superior returns and profitability. The reward for those organisations which are willing to rethink their risk management practices is the opportunity to generate greater wealth in a more socially responsible way by confidently and effectively managing high-risk investments in partnership with their external stakeholders.

Chapter 2

Risk and opportunity identification

2.1 Introduction

Writers on risk management have given far greater emphasis to the quantitative techniques of risk analysis than to the process of risk identification. Probability theory, Monte Carlo analysis, systems analysis, operations research, fuzzy set theories and decision theories are all examples of quantitative approaches that have dominated the risk management literature. While valuable in measuring objective risk levels, the neglect of the risk identification process is one of the reasons why many organisations often fail to manage their risks effectively. Ultimately, an unidentified risk cannot be managed, other than in a reactive way. For many organisations, a major step forward would be to be simply to get better at identifying risks. Once a risk has been detected, it ceases to be a surprise or an unexpected event, but instead becomes an explicit management challenge and therefore a potential opportunity. This exciting aspect of risk identification is often overlooked, the focus in most organisations being on problems rather than opportunities. It is not surprising therefore that the process is often seen as a gloomy affair, a perception exacerbated in the construction industry by confrontational contractual and procurement practices which create a culture of mistrust, blame, fear and conflict. The purpose of this chapter is to dispel this negative perception by exploring the risk identification process and discussing ways to help managers take better advantage of the risks and opportunities they face.

2.2 The role of contracts in risk and opportunity identification

The main mechanism for identifying and distributing risks and opportunities on projects are the contracts which bind parties together. Contracts are the legal mechanism by which project participants record their agreements regarding the distribution of risks and opportunities among them. These agreements are often standardised, although invariably some changes are always made to standard provisions which necessitate negotiations between the contracting parties. Consequently, a detailed understanding of the express and the implied terms contained in contracts is an important aspect of the risk and opportunity identification process. Express terms are those written in the contract and implied terms are those implied by the courts and by legislation.

While a detailed understanding of contracts and the law surrounding them is an important aspect of the risk and opportunity identification process, it is not the aim of this book to focus upon legal issues. There are numerous specialised and excellent books which already did this (Edwards 1995, Murdoch and Hughes 1996, Chappell 2001, Furmston 2003). Rather, the purpose of this chapter is to discuss the core principles of risk and opportunity identification which can be applied to a range of decision-making contexts. Although we do not ignore contractual and legislative issues, the danger of focussing on them exclusively is that managers may neglect those risks and opportunities for which they do not seem directly responsible, but which may affect them if not properly managed. Contracts have a detrimental effect on the sense of collective responsibility and open mindedness that is essential for the effective management of project risks and opportunities. They are no substitute for a comprehensive approach to risk and opportunity management but all too often they become so.

2.3   Understanding your objectives

The risk and opportunity identification process should commence while a decision is being made, rather than after it has been made, as is too often the case. Rather than starting with the identification of risks and opportunities, it should start with the identification of decision objectives because risks and opportunities are potential future events that can threaten or enhance them respectively. A decision's objectives provide the context within which the search for risks occurs and the concept of risk is meaningless without them. Unfortunately, many decisions are made automatically without a proper understanding of objectives which is one of the main reasons why many potential risks and opportunities are overlooked (Behling and Eckel 1991). Nevertheless, this can be easily avoided by following the five simple steps which are discussed in the following sections. The five steps are:

1   Obtain organisational commitment to risk and opportunity management.
2   Conduct a stakeholder analysis.
3   Consult stakeholders.
4   Identify objectives.
5   Identify key performance indicators (KPIs).

2.3.1   *Obtaining organisational commitment*

The foundation for effective risk and opportunity identification is genuine organisational commitment to the risk and opportunity management process. It is imperative that senior managers endorse and fully support a risk and opportunity-based management approach for their organisation. Without widespread commitment throughout an organisation, those involved in the risk and opportunity management process may not be fully committed and will not have access to the complete range of information they need to make effective decisions. Furthermore, those who have to act on those decisions will not understand or be committed to them.

The problem of securing organisational commitment can be particularly acute when a company is an independent subsidiary of a parent company. Problems can arise when the parent company holds important information about its risks, has different systems for managing them and has had little involvement in developing the risk management systems of the subsidiary. For example, one construction company which was an independent part of a larger mining group undertook an independent risk assessment of its construction operations at considerable expense, only to have this rejected by the parent group because it did not take into account important data which it held on operational risks and the first-rate technical experience of its global staff base. The mistake of the subsidiary was not to see its parent company as a key stakeholder in the risk management process. Consequently, they omitted important information and did not understand the end-user's needs and preferences.

2.3.2   *Conducting stakeholder analysis*

Any decision is made within the context of the wider goals, objectives and strategies of project stakeholders. As the notion of corporate responsibility gains increased priority, decision-makers are becoming increasingly aware of their social, moral and ethical responsibility to understand and accommodate the needs of stakeholders and to consult them at each stage of the risk management process. The importance of stakeholder commitment to policy and strategy formulation and implementation has been widely recognised (Burby 2001). If stakeholder commitment to risk control strategies is low, any plan, no matter how technically efficient, may be either ignored or half-heartedly implemented. In contrast, by involving stakeholders in an appropriate way, managers can build an informed constituency which has real commitment to take action to manage risks.

A stakeholder is any person or organisation which can affect, or can be affected by, the outcome of a decision (Freeman 1984). The variety of potential stakeholders associated with a project decision can be enormous and might include: contractors; sub-contractors; suppliers; consultants, the general public, regulatory, licensing and approval authorities and inspectors, unions; the client and its staff; operators; users and tenants of the final facility; the media; financial and insurance institutions; legal advisers; and special interest groups lobbying for issues such as minority or native rights, ecological, social and cultural sustainability, animal welfare etc. Effective consultation with stakeholders like these is critically important, to ensure that:

- Information for making decisions is maximised.
- Perceptions of risk and opportunities are the same among all stakeholders.
- Stakeholders understand the basis upon which a decision is made.
- Stakeholders feel involved and valued in the process.
- Stakeholders feel that their interests are being considered.
- Stakeholders understand their responsibilities and role in the risk management process.

The likely result of effective stakeholder consultation is that more risks and opportunities are identified, more ideas for managing them effectively are generated and greater long-term support for strategy to control them effectively is secured. Nevertheless, it is unfortunate that some industries such as construction have a poor record of stakeholder consultation (Preece *et al.* 1998). Although the reasons for this are unclear, it is likely to be related, in part, to the vast number of stakeholders that can be involved in activities such as construction projects. This makes widespread and meaningful consultation administratively difficult and a potential cause of heightened conflict rather than consensus. It may also be related to the difficulties that some stakeholders might have in participating effectively, due to physical distance, apathy, lack of resources and time or lack of skills to comprehend issues as complex as risk and opportunities. Other likely barriers to effective stakeholder consultation include lack of resources and time on the part of decision-makers to engage in meaningful consultation or lack of knowledge about how to do so effectively. Collectively, these factors create a significant gap between what theorists and educators claim is best practice and what actually occurs in reality.

Given the above constraints, it is not surprising that in the cases where stakeholder consultation has been undertaken seriously, the tendency has been to involve only those who demand involvement or are likely to support a decision. Understandably, without positive signals from constituents and under ever-increasing time and resource constraints, decision-makers in industries like construction are likely to see only obstacles and problems in securing their involvement. However, although time-consuming, the process of involving stakeholders, whatever their predisposition, is worth the effort and will benefit decision-makers enormously. In contrast, their exclusion from decision-making processes can lead to the neglect of important risks and opportunities. For example, this was vividly illustrated in the construction of the Millennium Rugby Stadium in Wales where poor communications between the Welsh Rugby Union and its neighbour Cardiff Rugby Club led the latter to refuse permission for tower cranes to swing over its air space. At significant cost to programme, budget and reputation, this led to a complete redesign of the masts that supported the stadium roof and to the recalculation of loads for the entire structure.

2.3.3   Consulting stakeholders

Burby (2001) found that effective stakeholder consultation involves a number of key decisions relating to:

- Objectives
- Timing
- Participants
- Techniques
- Information provision.

2.3.3.1   Objectives

Many stakeholder consultations are ineffective because the objectives driving the process are not clearly formulated. Possible objectives might include:

- Compliance with regulatory requirements.
- Giving stakeholders opportunities to voice opinions.
- Educating and informing stakeholders about hazards and opportunities and about strategies to manage them.
- Tapping stakeholder knowledge of hazards as a supplement to technical data.
- Understanding stakeholder perceptions and preferences to deal with hazards.
- Building a collaborative culture to mobilise a supportive constituency of stakeholders.
- Securing the participation and trust of stakeholders in the risk management process.

Given the increasing prominence of regulatory control in many industries and the increasing penalties for non-compliance, there is a danger that many managers will chose the first objective alone (compliance) as the motive for consultation. However, such a narrow approach is likely to result in significantly reduced benefits from the risk management process. For example, it could result in potential risks and opportunities going undiscovered, fewer ideas for dealing with those that are discovered and reduced commitment to management control strategies. Indeed, Burby (2001) found that when decision-makers pursued three or more of the above objectives, their constituents adopted 55 per cent more mitigation strategies than was the case in constituencies where only one objective was pursued. The greatest increase in stakeholder adoption of management strategies was when decision-makers emphasised "fostering stakeholder influence in hazard mitigation" as their main reason for consultation. Burby found a 76 per cent increase in adoption of mitigation strategies compared with decision-makers who did not have this objective. "Understanding stakeholder perceptions and preferences" was also important with a 70 per cent increase in strategy adoption.

2.3.3.2 Timing

Effective consultation requires effective planning and dedicated time. *Ad hoc* meetings attached to the end of other meetings send the wrong message to stakeholders, implying that the process is not meaningful and that their contributions are not valued. The result is a cynical atmosphere which is not conducive to constituency building. Clearly, decisions about timing are linked to the objectives underlying the consultation process. If the objective is simply compliance then meetings will be strictly dictated by laws and regulations. Similarly, if the objective is simply to inform stakeholders of a project rather than meaningfully gain their views, the consultation process might be limited to public hearings at the end of the project. However, if the objective is to tap stakeholder knowledge to assist in the identification of risks and opportunities and in the development of more effective control strategies, then the interaction will need to be far more frequent, occurring throughout the decision-making process. In particular, Burby (2001) found that large dividends in adoption of risk mitigation measures resulted from early stakeholder participation in decisions (85 per cent higher than managers who initiated stakeholder consultation in later stages of their decisions).

2.3.3.3 Participants

One problem in managing the consultation process is that, over time, the mix of stakeholders who need to be consulted may change. This means that consultative arrangements should be regularly monitored and reviewed. Furthermore, organisations cannot attend to all potential claims on their decisions and it is impossible and irrational, within the time and resource constraints of a project, to consult all stakeholders. Indeed, Burby (2001) found that decision-makers who indiscriminately embark on widespread consultations are only slightly more successful in seeing mitigation strategies acted upon. For these reasons, it is useful to employ a stakeholder management strategy, which can disentangle the important stakeholders from those which are less important. However, the literature on stakeholder management is enormous and bewildering and there are numerous models which are often difficult to operationalise. For simplicity, we suggest a model based on Freeman's (1984) classic definition of a stakeholder. This is illustrated in Figure 2.1 which classifies stakeholders as *Key, Important* or *Minor* according to their capacity to affect or be affected by a decision outcome. Key stakeholders should be a top priority and be intimately involved and consulted in the decision-making process. Important stakeholders should be a medium priority and involved to a lesser extent, being kept informed of the decision-making process. Minor stakeholders are a low priority and justify only minimal involvement in the decision-making process.

While this model is practically useful in effectively managing the stakeholder consultation process, it is rather clinical and lacks a sense of social inclusively and conscience. Depending on the project under consideration, the model will almost certainly lead to the exclusion of less powerful groups in society which are also disadvantaged in other ways. Without the consideration of minority groups, stakeholder consultations could easily become discussions among elites instead of meaningful and open discussions among multiple stakeholders. Unfortunately, this is often the case in practice. For example, Burby (2001) found that disadvantaged citizens only participate in 10 per cent of development planning decisions and that those advocating affordable social housing only participate in 1 per cent of development decisions. So there are ethical as well as commercial reasons for involving stakeholders in decision-making.

2.3.3.4 Techniques

The appropriate consultation technique depends on the objectives of the process. For example, if the intention is to simply inform the public of a

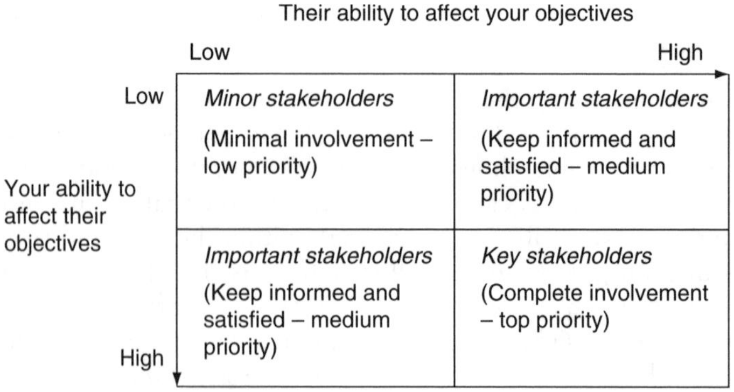

Figure 2.1 A simple stakeholder management model (adapted from Freeman 1984).

project, then public hearings are a widely used technique. Other common techniques include educational workshops, talks to community groups, brochures, newsletters, stakeholder advisory committees or working groups, expert panels, interviews, surveys and telephone hotlines etc. Each of these techniques is used for different reasons, results in different coverage of stakeholders and produces different types of information. For example, some stakeholders will not be, or are not, able to attend public meetings. Similarly, while surveys are useful for soliciting the views of a large number of people over a wide geographical area, they are not able to generate the same depth of understanding, insight and trust that can be achieved through interviews.

Stakeholder consultations should be at regular intervals or their timing could be dictated by specific decisions that need to be made (such as submitting a tender or not). The advantage of regular discussions is that they can often reveal new risks and opportunities as they arise, whereas periodic consultations which focus on specific decisions tend to be more narrowly defined.

Stakeholders can be consulted individually or in groups, the appropriate method depending on the significance of the decision being made and the practicalities of contacting stakeholders. While individual consultations may be appropriate for smaller decisions, the most effective consultation mechanism for decisions that involve significant risks and opportunities is to construct an expert panel (Bowden *et al.* 2001). Since the opinions of an expert panel will form the basis of the risk management process (actuarial data not being available for many risks and opportunities), the combination of skills and knowledge levels within the panel is an important consideration, as is the quality of facilitation. A well-conceived panel uses people who are trusted and respected, have the right mix of skills, experience and expertise, are well networked in their field and have the support of considerable resources and expertise from their own organisations. In many ways, it is the resources available to a panel rather than the individuals on it that is important. A panel with the right mix of knowledge and experience is unlikely to omit major events from consideration and is more likely to have its recommendations acted on. To this end, an expert panel should consist of people of high credibility and experience who genuinely understand a business and the range of potential risks deriving from its activities. When the stakes are high and decisions involve major risks, panels should consist of a combination of nationally and internationally recognised experts.

2.3.3.5 Information provision

Access to adequate and appropriate information is essential in order to empower stakeholders, to secure their involvement in the decision-making process and, ultimately, to ensure their acceptance of and commitment to any decisions made. For example, Burby (2001) found that the more the types of information provided by government planners in the development process, the more likely the recommendations proposed would be accepted by the community. Of particular importance in securing stakeholder acceptance was information relating to goals, hazards, alternative designs and risk mitigation strategies being considered. However, there are a number of potential problems which can starve the stakeholder consultation process of essential information. For example, some stakeholders may not have any risk-related information to contribute. This is highly likely in the construction industry where traditionally there is very little risk-related data collected and analysed – even by the largest organisations. Much risk information therefore remains anecdotal. Another problem in the construction industry relates to the inevitability that many stakeholders will have conflicting interests in a decision's outcome, adding a strong political/bargaining dimension to the decision of how much information to release to other stakeholders. There may also be confidential issues which one party may not wish to divulge. Indeed, even if this information was provided, some stakeholders might not have the expertise to make sense of it. The issue of information provision is therefore not just one of access and quantity but of content and context. Since different stakeholders are likely to understand information in different forms, it is therefore important that decision-makers disseminate information in a variety of ways.

2.3.4   Identifying and ranking stakeholder objectives

Every stakeholder will have a range of objectives that they want to achieve out of a decision. Some stakeholders will have the same objectives as the others and others will have very different objectives. Unfortunately, identifying them is not always easy. For example, some stakeholders will have never been asked to articulate their objectives before and others may not even know clearly what their objectives are. It is therefore likely that many stakeholders involved in a decision will have objectives which are imprecise, unrealistic and unsympathetic to the needs of other stakeholders. This can make it very difficult to make a decision without precipitating a conflict. To avoid this situation, it is useful to follow the few simple rules listed below:

- Make each stakeholder aware of your need to balance their objectives with other stakeholders' objectives.
- Make stakeholders feel comfortable and do not force them to adopt other stakeholders' value-structures.
- Assure stakeholders that they maintain some flexibility in their stated objectives.
- Make stakeholders feel important by valuing and accommodating their needs and opinions.

- Be sympathetic to the conflicting motives and pressures that stakeholders may be subjected to.
- Provide clear information about resource constraints that may make the attainment of some stakeholder objectives impossible.
- Ask as many "why" questions as possible. This prompts people to think through their objectives in more detail. Move from easy questions which are consistent with a stakeholder's value-structure to difficult ones that make them question these structures.

A further potential problem in eliciting stakeholder objectives is that people tend to express their objectives in ways that are not very useful in making a decision or measuring performance. For example, when asked for objectives relating to the construction of a new highway through natural forest, different stakeholders are likely to make a generalised range of statements such as: "minimise environmental damage", "minimise accidents", or "maximise profit". While these statements tell us that a stakeholder prefers more or less of something, they do not indicate the *relative desirability* of different levels of profit or environmental damage. Furthermore, they provide no clear definition of what is meant by environmental damage, an accident or profit etc. Therefore, it is important to encourage stakeholders to be more precise in their statements about objective. For example, a stakeholder might elaborate by saying that the objective is to "make a profit of $140,000". However, this is also problematic since it implies that any profit less than $140,000 is equally undesirable, which is unlikely to be the case. For instance, there is likely to be a point where the level of profit falls below that of other investment opportunities and makes the whole project unviable compared to other competing projects. Another stakeholder might elaborate by saying that one objective of the project is "to open up opportunities in a new market". However, this statement merely identifies a means to an end, which is probably to increase profits by a certain amount in a certain time period. To overcome these problems, it is important to ensure that stakeholders:

- Make precise rather than general statements of expectations.
- Identify ends not means.
- Break their objectives down into a *hierarchy*. For example, an objective such as "minimising environmental damage" might have lower level objectives that contribute to it, such as "minimising energy consumption" and "minimising the use of rainforest timber". A hierarchy provides a framework for decision-makers to understand, identify and manage the various components of risk involved in a decision.
- Rank objectives to reflect their relative importance. Ranking is important because it is impossible to accommodate all stakeholder objectives within the limited time and resources available to most decision-makers.

This will inevitably involve difficult trade-offs being made between different stakeholders.

2.3.5   Identifying key performance indicators (KPIs)

Having identified a ranked list of objectives in consultation with stakeholders, the next step is to determine how they will be measured by defining the KPIs associated with each objective. Since risk management is about achieving decision objectives, these *measurable* criteria then become the targets against which risk management success is measured and judged. Again, this must be done in consultation with stakeholders and it is important that the KPIs set are realistic and achievable. If they are not, then the project will inevitably be judged a failure and the risk management process will be a frustrating and impossible process. For example, in one petrochemical project in Australia, the client set a target of "zero accidents". A rudimentary risk analysis of the project made it clear that this target was unreasonable, impossible to achieve and of little value. Consequently, the target had to renegotiated before a more detailed and more achievable risk management plan could be developed.

However, few managers understand the critical difference between KPIs and objectives. The absence of appropriate KPIs can lead to a failure in risk management because without them it is impossible to determine when or whether the risk management process has been successful in achieving a decision-maker's objectives. The importance of KPIs is well illustrated by an example of one project in Sydney which involved the installation of a new filtering plant in a shipyard dry-dock. The objective of the project was to prevent toxic pollutants from ship cleaning activities spilling into Sydney harbour during the installation process. However, it was not until the project team began to ask how they would measure this that they realised that different stakeholders had very different ideas about what "pollution" meant in terms of the quantity and the nature of pollutants. Agreed KPIs eventually involved minute scientific measures of specific chemicals which helped focus the team's minds on what exactly they were trying to control. It also helped the team realise how difficult their task was, because the quantities involved were extremely small and there was very little margin for error in installing the equipment. It also helped them realise that special measurement processes and experts had to be employed as part of the risk management process.

While KPIs for objectives such as minimising pollution can be easily identified and quantified, KPIs for other objectives are less easily identifiable and quantified. For example, in the area of occupational health and safety (OHS), KPIs for measuring performance have been the subject of considerable and ongoing debate. For example, compared to "hard" objectives, such as cost and time, good OHS performance is often best measured in subjective terms. Furthermore, it is much more difficult to identify significant manifestations of organisational and other factors that could signal deterioration in safety performance before actual adverse safety impacts are realised. Finally, while it is easy to measure whether a project is a few days ahead of schedule or a few dollars under budget, good OHS performance is measured in "non-events" which are impossible to measure. For example, the most widely used measure of safety performance around the world is lost time injury frequency rates. This is measured by dividing the number of lost time injuries by the total number of hours worked on site per month. While it is an objective measure, many argue that this is a very poor indicator of safety performance, as it grossly oversimplifies the many intangible factors that collectively indicate how safe a site is. Also, these types of measures focus on the negative consequences of system failures and have been referred to as "negative" or "lag" indicators. However, it is also important to use "positive" or "lead" indicators of performance which show the extent to which system requirements are being met. These might include measures of performance such as "number of safety inspections", "number of people trained per month" or "appropriate resources available to allow managers to implement safety management systems" etc. In general lead indicators tend to be more intangible and difficult to measure than lag indicators, but are an equally important measure of performance.

Whatever measures are used, the end result of this stage of the risk management process should be a ranked list of KPIs that will help to prioritise and be an effective and meaningful benchmark for measuring the success of the risk management process.

2.4   Risk identification techniques

Only now, armed with a clear list of ranked objectives and KPIs, can a manager begin identifying the risks and opportunities associated with a decision. Ideally, all potential risks and opportunities should be proactively identified when the decision is being made so that they can be dealt with before they arise. However, while prevention is better than cure, no risk management system or any amount of diligence can identify all risks and opportunities in advance. This means that risk identification must have both a *reactive* and a *proactive* focus to be effective. In contrast to proactive risk identification, the purpose of reactive risk identification is to detect unforeseeable risks that arise after a decision has been made. These may have gone undetected due to inadequate proactive measures or arisen unexpectedly as a result of unpredictable events in the business environment or in workplace activities and processes. Unfortunately, most texts do not make a clear distinction between proactive and reactive risk identification techniques. Consequently, many companies become confused in what they are trying to achieve when managing risks and opportunities, or even miss some risks and opportunities completely. To avoid this, the following sections treat separately the various techniques of proactive and reactive risk identification.

2.5   Proactive risk identification techniques

In essence, proactive risk identification involves *imagining* potential *future events* that could affect the attainment of decision objectives, either negatively or positively. This is not easy because, in most cases, personal interests ensure that people's assumptions about future events are biased in favour of their own interests and experiences. Herein lies the challenge of risk identification, which is to question the most fundamental beliefs, values and desires which constrain our thinking so that we can go beyond what we *want* to happen or *expect* to happen to what *could* happen. This takes open mindedness, courage and creativity.

Managers can do much to enhance the creative potential of an organisation. The main strategies are to recruit people with creative abilities, train people to work more creatively, create an organisational culture and structure which is conducive to creativity and use idea-elicitation techniques to help people think more creatively. Collectively, these methods represent a complete strategy to facilitate proactive risk identification within an organisation. Nevertheless, the limited literature in the area of risk identification has almost entirely focussed upon classic idea-elicitation techniques such as brainstorming. Clearly, such techniques have an important role to play in the risk identification process but it is also important that managers adopt a more imaginative and comprehensive strategy to foster creativity.

2.5.1   Employing and using creative people

The most obvious way to increase an organisation's creative abilities is simply to employ more creative people or use creative employees more effectively. However, it is not easy to identify creative people. For example, since creative acts demand special mental abilities we might expect that creative people have especially high IQs. However, while psychologists have found that low IQ people do tend to be less creative than high IQ people, the relationship is very weak (Szilagyi and Wallace 1987, Belbin 1997). Furthermore, Belbin (1997) found that groups of "clever people" generally performed badly, being characterised by destructive debate, intolerance and a lack of coherence. It is also interesting to note that many of history's most creative people, such as Copernicus, Rembrandt and Faraday, are estimated to have had IQs of less than 110.

Instead of depending on IQ, there seems to be general consensus that creative acts rely on many mental processes working together in harmony, namely problem finding, idea generation, imagination, simplification, risk taking and motivation to learn (Santrock 1998). That is, creative people excel at finding problems, at finding new perspectives in their solution and at producing order out of chaos. They are also willing to take risks, to learn from failure and are determined, unconventional, self-confident, tolerant of ambiguity and intrinsically rather than extrinsically motivated by things such as money, evaluations, prizes etc. For example, one of Albert Einstein's main motivations in life was to find simplicity and to disseminate his ideas without undue publicity. Indeed, when asked why he used hand soap for shaving instead of shaving cream, he replied that using one bar of soap was less complicated. It is worth noting that a number of psychological tests have been developed to identify creative individuals, some based upon personality measures, some on biographical experiences, some on intellect and others on cognitive skills such as divergent thinking. However, there is controversy surrounding these tests and only inconsistent evidence of their ability to predict real world creative achievement (Wallach 1985). Consequently, contemporary creativity tests focus upon people's *outputs* rather than upon their mental abilities (Amabile 1983). For example, in rating Frank Lloyd Wright as one of the most creative architects of his day, one would tend to cite as evidence, his buildings rather than his personality traits, although, admittedly, he often behaved and wrote eccentrically (Lawson 1990).

Unfortunately, creative individuals like Einstein and Frank Lloyd Wright are rare and most managers need to elicit creative potential from specific combinations of relatively uncreative individuals. Belbin's (1997) research has been most influential in determining which combinations of roles, capabilities and personalities induce creative tensions in teams. Nevertheless, it remains the case that most management teams are created in haste without proper regard to such issues.

2.5.2   Creativity training

An alternative to recruiting creative people is to train an existing workforce to be more creative. Most creativity training programmes are based upon the creative problem-solving (CPS) programme developed by Sidney Parnes (Parnes *et al*. 1977). This programme teaches participants a range of techniques to help them find facts, problems, ideas, solutions and overcome resistance to their implementation. Unfortunately, while creativity training does seem to produce changes in creative ability, the effect is often short-lived (Basadur *et al*. 1986). Therefore, until there have been more long-term studies of its effectiveness, its value to managers remains uncertain and it needs reinforcing with other approaches to learning which can open people's minds to potential risks and opportunities.

One learning approach that could facilitate greater risk awareness in projects is *project closure reviews* (Royer 2000). This simply involves spending some time at the end of a project reflecting upon the lessons learned and transferring them to people in future projects. For example, Ernst and Young use this technique by constructing a risk database using a standard set of questionnaires that are completed at the end of every project by different stakeholders. Future project managers can access this database to predict possible risks and develop effective management strategies, based on past company experience. In effect, this is a process of knowledge management. Similarly, British Petroleum has created a post-project appraisal unit with the sole mission of helping the company learn from its mistakes and successes. The unit's objective is to improve company performance and help managers formulate investment decisions more accurately, appraise them more objectively and execute them more effectively (Gulliver 1991).

2.5.3   Organisational characteristics

For recruitment and training strategies to work, the structural and cultural aspects of an organisation must also be conducive to creativity. The main organisational characteristics which affect creativity are leadership, organisational structure, organisational climate and culture, and environmental relationships (King and Anderson 1995).

2.5.3.1   Leadership

By the 1980s consensus had emerged about the leadership styles that encourage creativity. These are styles that allow people to participate in decision-making and to suggest novel ideas without fear of blame or recrimination. A second vital attribute is the ability of leaders to provide a vision to which people can commit. It is interesting to note that this type of leadership need not come from people with formal leadership authority and responsibility. Indeed, throughout history one tends to find that those who have developed great ideas are more often than not "champions" who feel a strong personal commitment to an idea than formally prescribed leaders who have been given responsibility to develop it. Furthermore, the people who generate the best ideas are often as likely to be technicians as senior managers.

2.5.3.2   Organisational structure

It is widely accepted that *organic* rather than *mechanistic* structures are the best suited for creativity. The important distinction between these organisational structures was first drawn by Burns and Stalker (1961) whose work, which is summarised in Table 2.1, continues to influence our understanding of how organisations work.

Table 2.1 The characteristics of organic and mechanistic structures

Mechanistic structures	Organic structures
• Centralised decision-making.	• Decentralised decision-making.
• Hierarchical structure with stable divisions/departments based strongly around functions.	• Flat structure with temporary work groups/teams based around specific projects.
• Vertical communications dominate.	• Lateral communications dominate.
• Formal communications dominate.	• Informal communications permitted.
• Rigid job definitions set by senior managers resulting in a high level of specialisation.	• Flexible job definitions defined by individuals through interaction and negotiation with colleagues who decide the best way to get the job done.
• Many rigid rules and procedures which result in prescribed work patterns.	• Few rigid rules and procedures.
• Power and authority based upon formal seniority within hierarchy.	• Power and authority changing with changing circumstances and based upon individual skills and abilities.

Source: Adapted from Burns and Stalker (1961).

In terms of creativity, the advantage of the organic structure is that it provides people with a relatively high degree of autonomy and control over their jobs, giving them the confidence to experiment with new ideas and to interact with others in their development, without fear of failure. However, more recent research has elaborated our understanding of how organisational structure fosters creativity. For example, we have discovered that while organic structures are best for the development of new ideas, mechanistic structures are best during the implementation phase of the creative process, because new ideas often require backing from sources of authority to overcome resistance (King 1992). This has important implications for risk management practices because it indicates that while organic structures might be best for identifying risks, mechanistic structures might be best for their subsequent management.

Another interesting research has been conducted by Bookstaber (1999) using lessons of survival from the natural world to provide new insights into the cognitive processes by which people identify and manage risks and opportunities. He noted that any animal that has found a well-defined niche tends to follow a specialised rule for detecting risks that depends critically upon that animal's perception of the world. If the world continues to exist as the animal perceives it, with the same predators, food sources and landscape then the animal will survive. However, if the world changes in ways beyond the animal's experiences then, unless it can adapt, its survival may be threatened. This is a warning to firms who have come to rely on niche markets because it indicates that precision in addressing the known often comes at the cost of reduced flexibility to address the unknown. It also has important implications for managers of complex projects. First, it advocates moving resources away from dealing with detailed investigations of known risks towards risk management structures with more sensitivity to unknown risks. Second, it implies reducing organisational complexity and the hierarchy of responsibility for detecting risks. Complex and hierarchical risk management systems are likely to obscure unidentified risks that fall across organisational boundaries and slow an organisation's responsiveness to events beyond the intended design and function of that system. This is supported by the analysis of many disasters throughout history which have revealed that the root problem in most cases is not the complexity of the unseen risk which normally turns out to be simple, but the complexity of the organisation. For example, the near melt-down of a nuclear reactor on America's Three Mile Island in the 1960s was attributed to a safety system which was so complex that it became incomprehensible. This was also the case in the Occidental Piper Alpha disaster on 6 July 1988, where the long chain of events that led to the eventual fire were initiated by a simple unfinished maintenance job in the gas compression module. Again, safety procedures were so complex that they exacerbated the problem rather than help to resolve it. The lesson for managers is clear. They should reduce complexity and trade-off the pinpoint monitoring of known risks in favour of a lower resolution and broader band monitoring of unknown risks. Specialised monitoring systems developed in response to what is known can only provide a perspective on the same whereas "coarse measures" will be more likely to identify areas of unanticipated risk.

2.5.3.3   Organisational climate and culture

Most managers are familiar with the idea of organisational culture. This refers to the values, norms, beliefs and assumptions embraced by an organisation's participants. However, a less familiar concept is that of organisational climate which refers to the behaviours which characterise organisational life and which are perceived to be acceptable by its members (Nystrom 1990). Recommendations regarding creative climates include being open to change, encouraging and supporting of risk taking, and being tolerant of vigorous debate, freedom of thought and a playful approach to new ideas. The recommendation for creative cultures is to emphasise individual autonomy, egalitarianism, informality, mutual accountability, collective responsibility, flexibility, adaptability and, surprisingly, disharmony. Conflict is seen as a constructive force for creativity if it is managed well. However, what is often missing is the ability to manage it constructively, which is the prerequisite to it being a creative rather than destructive force (Loosemore *et al.* 2000).

2.5.3.4   Environmental interaction

The ability of people to identify business risks is influenced by the way they interact with an organisation's environment. A tragic example of a disaster which was partly caused by an organisation's ignoring its environment was the Aberfan coal slip disaster in South Wales which killed 144 people. Here, a sense of organisational exclusivity developed within the Coal Board which led to a sense of superiority over non-members. In Aberfan, the local council foresaw the disaster but were labelled as "cranks" and repeatedly "fobbed-off" with ambiguous and misleading statements from the Coal Board such as "we are constantly checking these tips". As Turner and Pigeon's (1997: 49) analysis of the disaster showed, there was "an attitude that those in the organization knew better than outsiders about the hazards of the situation with which they were dealing". For this reason, it is critical that project managers maintain regular contact with a project's environment and utilise the extensive knowledge which resides in it, to identify potential risks and opportunities.

2.5.4   Idea elicitation techniques

Idea elicitation techniques help individuals structure their thinking so that they more fully understand the risks and opportunities associated with a decision. As René Descartes noted, "*It is not enough to have a good mind. The main thing is to use it well*". These techniques complement the methods discussed above which are designed to create the right environment in which they can be effectively employed. There is a range of idea elicitation techniques appropriate to different situations and these are discussed below, in order of increasing complexity.

2.5.4.1   Checklists

The simplest way to identify risks and opportunities is to use a checklist of them compiled from experiences on previous projects. A standard checklist of risks and opportunities that can typically affect a construction project's objectives is illustrated in Appendix B. It is not an exhaustive list and should only be used as a starting point to develop and continually update one's own checklist, based upon one's own experiences. Nevertheless, having created broad risk categories, it is possible to continue breaking categories into sub-categories until the array of risks facing a project has been rigorously dissected and defined. For example, the *chemical spill* sub-category in the general *technology* category in Appendix B can be further split into sub-subcategories which would demand different management strategies such as *toxic liquid escape, toxic gas escape, toxic smoke escape, toxic radioactive escape* etc.

While the list of risks and opportunities in Appendix B is useful, the generic method of categorisation might not suit the needs of every organisation. For example, there is no indication of when or where these different risks and opportunities might arise during the course of a project. For this reason, an organisation might choose to adapt the checklist to its own specific business. For example, the list could be reorganised according to:

- the level of decision-making affected (corporate/strategic, commercial/financial/tactical, operational/technical/project)
- the stage of a project they arise in (feasibility, concept, design, tendering, construction, handover and transition, operation etc)
- the impact on organisational goals (safety, financial, time, quality and environmental).

The ease of the checklist approach has made it the most common approach in the construction industry and a number of the so-called "expert systems" rely upon them. However, checklists are often used inappropriately and the following warnings should be heeded:

- Take great care in transferring risk checklists between different decisions or projects. It is highly unlikely that any prior checklist will comprehensively cover all risks associated with a particular decision or project.
- Checklists are very useful when there is a wealth of experience from the past. This means that they are most suited to standard and routine decisions. However, they can constrain thinking when making non-routine decisions that have not arisen before.
- Checklists are merely memory prompts that should encourage further investigations into the potential risks or opportunities identified. They do not help determine *how* and *when* risks could occur. They only help to identify *what* threats or opportunities could occur.
- Checklists treat risks as isolated events and fail to consider that most risks arise from a chain of interdependent activities that combine to collectively produce a certain type of impact or consequence.

2.5.4.2   Decomposition techniques

The simple task of planning a decision or project and breaking it down into its component parts can help identify the potential risks and opportunities involved. This is a common method of risk identification used in construction projects where "work breakdown statements" or "method statements" are often used to divide an activity into a simple set of steps, operations or activities which can be analysed in isolation. It is a valuable process that encourages a decision-maker to think through a decision in a logical, incremental and structured way and provides a useful audit trail for future risk management activities.

In creating a work breakdown statement, stakeholders should be consulted by forming a committee or discussion group. Working alone is always a mistake and will inevitably result in a narrow range of risks and opportunities being identified. However, while the process of breaking down a process into parts is useful, it does not replace the need to be creative in identifying potential risks and opportunities. Therefore, it is worthwhile using a technique that can help a group generate ideas about potential risks and opportunities. A number of techniques to help do this are discussed below.

THE DEVIL'S ADVOCATE

Devil's advocates are independent people introduced into a group to formally introduce dissent into the decision-making process and to question assumptions underlying specific decisions. They provide a different perspective and encourage a re-analysis of a problem that can highlight hitherto unforseen and unacceptable risks (Schenk 1984). Effective devil's advocates work through constructive conflict and need to be trained to avoid the process becoming destructive. Furthermore, the personality, experience and credibility of the person selected is critical in securing the respect of group members. A famous example of excellent devil's advocacy was Robert F. Kennedy's involvement in the Cuban Missile Crisis in 1962 which prompted a re-analysis of the problem, highlighting unforseen risks which almost certainly averted a nuclear war.

SCENARIO BUILDING

Scenario building involves speculating about the future. One way of building scenarios is to ask "*if then*" questions about specific events such as, "*if oil prices doubled, then...*". Another way of building scenarios is asking "*what if*" questions such as "*what if the client went bankrupt?*" One possibility in framing such questions is to consider what is the norm and ask what would happen if the norm did not occur. This approach prompts people to think laterally and to imagine how things can go worse or better than expected.

ATTRIBUTE LISTING

Attribute listing involves identifying as many attributes with the decision object as possible and using these to generate ideas about possible risks. For example, in considering the risks associated with operating a hoist, one would list the attributes of a hoist such as fast, heavy, clumsy, sharp, dirty, noisy etc. By considering these attributes, risks become more obvious. For example, noisy would suggest possible hearing damage to users, dirty might indicate potential dermatological risks etc.

FORCED RELATIONSHIPS

This involves finding ways in which normally unrelated ideas or objects may be related. For instance, when looking at the risks associated with working on an inner-city construction project one may force together the objects of "crane" and "adjacent building". This would encourage one to think about how the crane could swing over adjacent building air space, collide with the adjacent building, drop things on the adjacent building, create noise that interferes with those who work or live in the adjacent building etc. Start this process by listing the components or objects associated with a decision and other related decisions and then randomly combine them to look for possible risky or opportunistic relationships.

SYNECTICS

Synectics relies upon the use of analogy and metaphor to generate new ideas. Through metaphors, decision-makers are able to reconstruct what they already know in new ways and to connect seemingly unrelated ideas, objects and processes. For example, the Canon mini-photocopier contains a unique throw-away print drum which was modelled on a beer can. Before this idea, the print drum accounted for 90 per cent of maintenance problems and the breakthrough came when the task-force leader ordered some beer at a late-night meeting, stimulating one of the team to question whether the same process for making an aluminium can could be used to make a copier drum (Nonaka 1991). To prompt this process, try to draw analogies between your current problem and similar problems in other contexts. The trick is to let your imagination run wild and think outside the square.

2.5.4.3   Forecasting

Forecasting is widely used to identify quantifiable risks and opportunities. It involves analysing and evaluating past information and statistically using the results to predict future trends. A good example of forecasting is the life cycle costing of building components using discounted cash flow (DCF) techniques.

There are three main types of forecasting methods:

1   Extrapolative forecasts – based upon the belief that history repeats itself.
2   Causal forecasts – based on using cause-and-effect relationships to predict the future.
3   Normative forecasts – assume that people take an active role in shaping the future and try to take their goals and values into account when predicting it.

While forecasting is useful for identifying trends, it has a number of limitations:

- It is totally dependent upon the quality of historical data and the records kept by an organisation. Haphazard record keeping means haphazard forecasts.
- It is only useful for identifying *trends*. It is not useful for identifying specific events.
- It is only useful for extrapolating things that happened before. It cannot be used to explore unique problems that have never happened before.
- It is of little use when considering risks that are resistant to measurement.
- It is widely accepted that the numerous assumptions made by managers in making forecasts can bias predictions, resulting in a self-fulfilling prophecy.

2.5.4.4   Soft systems analysis

Soft systems analysis can assist in identifying the feelings, attitudes and perceptions of stakeholders, enabling potential conflicts between them to be considered (Stewart and Fortune 1995). This technique involves the construction of *rich pictures* that represent holistic pictorial representations of:

- key players in a project
- where resources come from
- where constraints might emanate from.

An example of a rich picture based on the construction of the Sydney 2000 Olympic games Stadium in Australia is illustrated in Figure 2.2. It identifies potential risks relating to: finance (funding, inflation, investment returns and construction costs), physical conditions (geology and weather), stakeholders (government, public and local authorities) and relationships with suppliers, construction teams, health and safety inspectors etc.

An additional way of looking at this project is to construct a *systems map* which represents a conceptual snapshot of the main internal and external components of a problem under investigation. This is illustrated in Figure 2.3.

By inserting relationships between components within a system map, a *systems influence diagram* can be produced resulting in a greater understanding of risk dynamics and of further sources of potential risks and opportunities in the relationships themselves. This is illustrated in Figure 2.4.

2.5.4.5   Brainstorming

Brainstorming is a group-based process which is valuable when making decisions about new, large, complex and non-standard business activities. It was developed by Osborn (1953) and relies upon group dynamics to elicit ideas. The success of brainstorming depends upon the breadth of

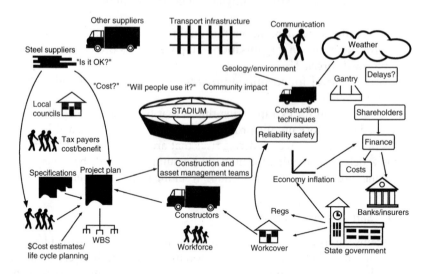

Figure 2.2 Rich picture of Sydney 2000 Olympic Stadium (adapted from Stewart and Fortune 1995).

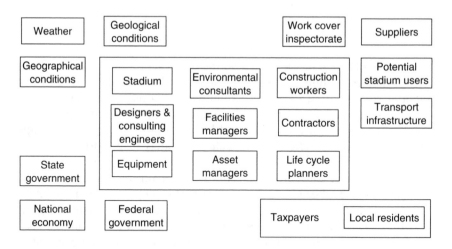

Figure 2.3 Systems map of Sydney Olympic Stadium construction (adapted from Stewart and Fortune 1995).

experiences and perspectives within the brainstorming group and the skills of the facilitator in combining them effectively. It also requires careful planning because of the time and resources involved.

A typical brainstorming group consists of between 10 and 15 stakeholders in a decision. Ideally, the group should draw from different disciplines who

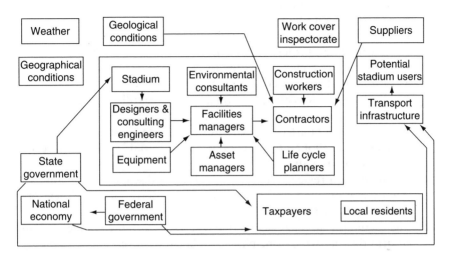

Figure 2.4 System influence map of Sydney Olympic Stadium construction (adapted from Stewart and Fortune 1995).

are key members of a project team and involve any specialists who can bring additional expertise to the process. It is critical to have a range of interests represented to prevent polarisation of views, and the inclusion of stakeholders encourages collective responsibility for the identification of risks and opportunities. In particular, when external stakeholders are involved, there are also numerous public relations benefits and the group will gain enormously from an outsider's perspective.

Brainstorming is a structured process that should involve the following steps:

1 Identify a facilitator who will present the problem and manage the process. Ideally, the facilitator should be an independent and experienced person with no interest in the decision outcome.
2 Identify a recording secretary who will record ideas but not contribute.
3 Select the brainstorming team using the above criteria and brief them in advance about the purpose of the exercise and expected outcome. They should come prepared.
4 The facilitator begins by reviewing the decision to be made, the purpose of the workshop and its structure.
5 The member most familiar with the decision defines it so that everyone understands what is being considered.
6 The first phase of risk identification lasts about 30 minutes and involves each group member writing his or her ideas down on paper. The facilitator should ensure that the principles guiding this stage are:

   • Let your imagination run wild.
   • Outlandish suggestions are encouraged.

- Generate as many ideas as possible.
- Quantity not quality.
- Anything goes.
- There are no professional territories.
- Anyone can comment on any issue.
- People's ideas are kept anonymous.
- Everyone's ideas are given equal weight regardless of seniority.

7   The ideas are submitted anonymously and the facilitator lists them on a whiteboard without any discussion. Ideas should not be associated with specific individuals.
8   The team then reviews the list through open discussion. The aim is to classify and group similar risks and where appropriate, add new ones as ideas are generated.
9   The facilitator summarises the ideas, lists them on a whiteboard and ensures the following principles are adhered to during discussion:

- Only one person speaks at a time.
- Everyone must have an equal chance to participate.
- No one member should be allowed to dominate.
- People are encouraged to combine ideas and build on those suggested by other members.
- Ideas should be evaluated without criticism.
- There are no professional territories.
- Anyone can comment on any issue.

10 The group is then reduced to five key members who have the greatest stake in the outcome. Their task is to simplify the list if needs be, and produce a list of about 10–15 risks. However, this will vary from one type of decision to the next and it is better to have too many risks than too few. Risks should not be omitted just to keep the list short. Discarded risks should also be recorded to maintain an audit trail and facilitate later review if necessary.

Unfortunately, while groups can be very creative, there are a number of factors which can reduce their effectiveness in identifying risks and opportunities. For example, in most projects there are a range of conflicting interests which are likely to reduce communication efficiency within a group. Conversely, in highly cohesive groups, there might be problems of social influence that act to suppress creative ideas. This is known as the "Asch Effect" (Asch 1956) where people's thoughts become polarised in the direction of the shared values of the group or around the views of one individual dominant member. Another harmful effect associated with cohesive groups is "Groupthink" (Janis 1988).

This occurs when group members become overly protective of each other and hide risks that could potentially threaten or expose group members to recrimination or blame.

For these reasons, brainstorming is often managed using methods such as the *Delphi Technique* and the *Nominal Group Technique*. Alternatively, some organisations prefer to avoid the use of groups altogether by relying on the opinions of individual risk analysts working by themselves or in one-to-one interviews with organisational members. Nevertheless, groups remain the most common medium of risk identification and for this reason it is worthwhile considering the *Delphi Technique* and the *Nominal Group Technique* in more detail.

THE DELPHI TECHNIQUE

The Delphi technique involves a coordinator carefully constructing a series of small multi-disciplinary problem-solving groups to discuss potential opportunities and problems which could influence the outcome of a decision. The difference between Delphi groups and brainstorming groups is that the former never physically meet but are usually coordinated through e-mail, the worldwide web or in writing. The Asch Effect and the Groupthink effect are therefore minimised due to the lack of interaction between respondents.

The Delphi process starts with the coordinator asking group members on an individual basis for their opinions about a certain problem, usually in some form of questionnaire. After a pre-specified period, ideas are returned and anonymously summarised by the coordinator and re-distributed for further discussion. The important point about this initial session is that outlandish suggestions are encouraged and people are not restricted to their own knowledge domain. Furthermore, ideas are not associated with specific individuals.

A second stage of opinions are then sought which are quite different from stage one in that they are more evaluative and based on the ideas generated from stage one. Once again the coordinator summarises the ideas and after a number of further rounds of discussion when opinions have stabilised, a final consensus list is produced which is a team output rather than an individual one.

While overcoming some of the problems associated with traditional brain-storming, there are a number of weaknesses with the Delphi technique. For example, people tend to find out who is in the Delphi group and might exert pressure upon each other, beyond a manager's control. Furthermore, it is difficult to ensure the diligence of participants and to maintain their motivation to contribute. Finally, the process takes considerable preparation, is often slow and depends upon the respondents' abilities to express themselves clearly in writing.

THE NOMINAL GROUP TECHNIQUE

The nominal group technique is another variant of brainstorming. It commences with a group of between seven to ten members who are separated from their normal work context and asked to generate ideas relating to the solution of a problem. This is usually done first on paper, without any discussion. Then each individual briefly presents one of his or her ideas which are recorded onto a flip chart. This process continues until all ideas are exhausted and then discussion commences around each idea in turn, everyone contributing equally with the intention of clarifying and evaluating each idea in terms of the risks posed. The focus must be on the ideas rather than the individuals who generated them and the process should be managed in a positive manner rather than being negative and critical. Eventually, at the end of the discussion, during which people can take notes, each individual ranks the various risks in order of seriousness and then they are mathematically aggregated to find a group decision.

While overcoming some of the shortfalls of traditional brainstorming, the main problem with this method is the inflexibility of the highly structured approach. Furthermore, strong management is needed to prevent dominant personalities controlling the discussions (Chapman 1998). Finally, the structure of these groups is critical. In particular, diversity is important since group cohesiveness and homogeneity are strong correlates of the Asch Effect. This is not as straightforward as it seems and involves considering the knowledge, experience and personalities of potential members. For example, while it might seem logical to include experts, they can be destructive since their imagination is often constrained by their experiences. Furthermore, experts tend to have high levels of self-belief which can make them seem infallible and irreproachable.

2.5.4.6   Electronic brainstorming

On many large projects, where team members are spread around the country or world, it is difficult to bring team members together for a brainstorming session. IT developments are making this more feasible and there is research to indicate that electronic brainstorming can overcome many of the problems associated with traditional brainstorming techniques (LaPlante 1998).

There is now a range of specialist software packages available to facilitate electronic brainstorming but e-mail can be just as effective. Typically, the process starts with the facilitator e-mailing group members on an individual basis for their opinions about a certain decision. Any documentation needed can be electronically attached to the request. The important point about this initial session is that outlandish suggestions are encouraged and people are not restricted to their own knowledge domain. After a pre-specified period, ideas are returned and anonymously summarised by the facilitator and re-distributed for further discussion. Ideas are not associated with specific individuals and a second stage of opinions is then sought which is quite different from stage one in that the elicited opinions are now more evaluative. Once again the coordinator summarises the ideas and after a number of further rounds of discussion, when opinions have stabilised, a final consensus list is produced.

Electronic brainstorming need not be restricted to decisions with geographically dispersed stakeholders. However, it needs careful management for the following reasons:

- Group members may secretly interact and exert pressure upon each other.
- It is difficult to ensure the diligence of participants in contributing.
- The process takes considerable preparation.
- The process can be slow.
- The process depends upon people's abilities to express themselves clearly in writing.
- The process depends upon people's abilities to use e-mail.
- It restricts the group dynamics that are so valuable to the creative process.

2.5.4.7 Influence diagrams

Influence diagrams (sometimes called "Ishikawa diagrams" or "tree diagrams") can be used to help you discover *how* a threat or opportunity might arise. It should be used only after you have used another technique to identify *what* threats and opportunities might occur and *when* they may occur. Such techniques include checklists, method statements, forecasting, soft systems analysis and brainstorming.

Influence diagrams recognise that most threats and opportunities do not occur in isolation but arise from a chain of contributory events (or sub-risks). As Blockley (1996) noted, most things have one thing wrong with them and, very often, it only takes a second fault to make it a problem. A threat or opportunity cannot be understood fully in isolation from this interdependency and the effective management of a potential threat or opportunity is only possible when you understand this whole process. To this end, influence diagrams are used as a graphical representation of the chain of contributory events which could lead to a risk or opportunity eventuating. An example is provided in Figure 2.5, which reveals the component events that could cause a cost overrun to arise (the horizontal arrow) and how they could combine to do so.

The process of constructing an influence diagram is simple and involves dividing the main threat or opportunity into its components and subcomponents until the origins of a risk are identified. This process can be facilitated by working backwards from the eventual threat or opportunity, asking *"what could cause?"* questions. In the above example relating to a

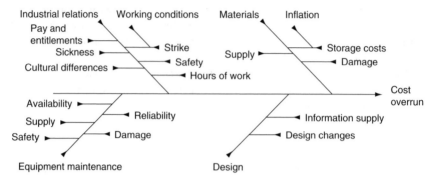

Figure 2.5 A typical influence diagram (adapted from Shen 1999).

cost target failure, the chain of events depicted would have evolved from asking the following questions:

1 What could cause a cost target failure?
   Answer – An escalation in the costs of materials, labour, plant or a design disturbance etc.
2 What could cause escalation in the costs of materials, labour, plant or a design disturbance?
   Answer (for materials) – Lost in store, damage, inflation.
3 What could cause materials to be lost in store, damaged, inflation?
   Answer (for lost in store) – Security problems.

2.5.4.8 Fault Tree Analysis

A sophisticated analytical technique which uses tree diagrams to predict risks is *Fault Tree Analysis*. This was originally developed in the US to prevent the accidental launching of missiles and has been used extensively in safety engineering ever since (Cameron 1984). Fault Tree analysis involves looking for potential faults or weaknesses in a system that might cause failure and mapping the connections between them. Fault trees can be used to help you discover *how* a threat or opportunity might arise. It should be used only after you have used another approach to identify *what* threats and opportunities might occur and *when* they may occur.

To construct a fault tree, follow the steps below:

1 Identify "*top faults (risks)*" – a major problem that could exist to prevent the attainment of decision objectives.

2   Identify *"secondary faults"* which could contribute to the occurrence of the top fault. These compose the top branches of the fault tree.

3   Identify the *"primary faults"* which could contribute to the occurrence of the secondary faults.

4   The process continues until all potential faults have been exhausted. This might stop anywhere between 4 and 16 levels and faults on lower levels of the tree are more specific than those at the top.

A typical fault tree is illustrated in Figure 2.6 with symbols indicating the logic through which different faults are interconnected.

2.5.4.9   Simulation

Simulation uses mathematical modelling techniques to help managers artificially experience a situation and thereby identify the potential risks and opportunities associated with it. The advantage of simulation over the largely manual techniques reviewed in previous sections is its ability to handle huge quantities of information and to take into account the interdependence between different risk variables. That is, how one risk can create another and how a particular combination of circumstances can impact upon a project variable. Simulations also allow managers to experiment by altering project variables to see what the impact on various risk levels will be.

Computers are essential to undertake this process where the computer acts as an experimental laboratory where the project can be "run" over-and-over again using different combinations of input assumptions. Such is the sophistication and increasing availability of simulation tools to the general public, that the US navy has started to issue its pilots with a modified version of Microsoft's "Flight Simulator" after Cadets who had used it, scored higher scores in flight training than those who have used traditional flight simulators. The business world has noted the potential value of simulating the business environment in much the same way as one can model an aircraft. For example, Powerslim in the US produced a simulation package that enables people to evaluate the potential benefits and risks of expanding into new markets or investing in new equipment. In construction, the AROUSAL training package developed by Professor Peter Lansley at The University of Reading in the UK does much the same thing. The latest development in this field is to link real data from a company's accounts and records so that managers can ask "what if" questions about the future.

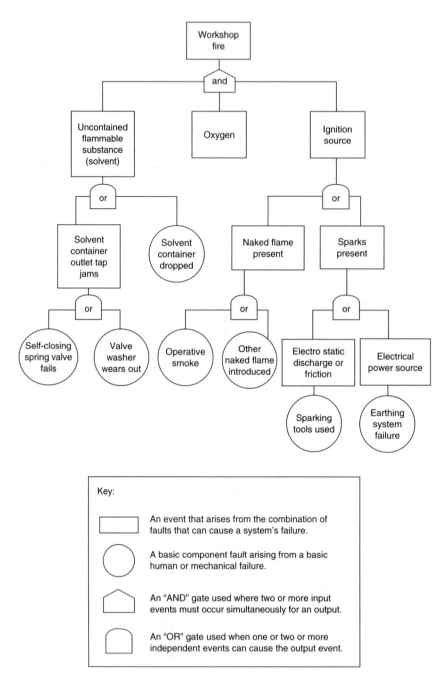

Figure 2.6 A typical fault tree (Loosemore *et al.* 1999).

2.6   Reactive risk identification techniques

No matter how rigorously a manager applies proactive risk identification techniques, it is inevitable that some risks and opportunities will arise after a decision has been made or after a project has moved forward and is progressing through its implementation stages. It is impossible to identify all potential risks and opportunities in advance and new risks and opportunities often arise as a result of completely unpredictable events. There is therefore a need to have the capacity to effectively and efficiently react to these risks and opportunities when they arise. This can be done by simply encouraging employees to notify their supervisor of potential hazards, when they become aware of them. However, there are also a number of formal techniques that can assist in this process. These are discussed below.

2.6.1   Risk inspections

Risk inspections involve inspecting the workplace, employees and/or documents at regular intervals. The aim is to identify new potential threats and opportunities to a decision outcome that arise while it is being implemented. Regular inspections are particularly important when documents or the workplace is changing continuously. However, it is important that they are not seen as something:

- That are restricted to site activities since they are equally useful in the office environment.
- That are restricted to physical risks since they are equally useful for identifying those which may be contained within documents.

Formal risk inspections should be made regularly until a decision is fully implemented or project is completed and planned outcomes are achieved. Where necessary, stakeholders and specialists who can contribute to the identification of potential risks and opportunities should also be involved. For example, when toxic risks exist, doctors may be involved to monitor employees' health through medical testing.

2.6.2   Bug listing

The idea behind bug listing is to list things that tend to bother people on a day-to-day basis, which could potentially interfere with decision outcomes. The problem with bugs is that they are so commonplace that they are hard to remember. People tend to miss them by focussing on the larger problems that might arise in their formal inspections. However, the cumulative effect of minor problems can be very important.

Bug lists are best made by carrying a notebook to record the bugs when they arise. These lists can form the basis of risk identification in regular risk review meetings (see Section 2.6.3). A list of bugs might be as simple as the following:

- Untidy work areas.
- People being continually late with information.
- The gate that does not close properly.
- People who do not wear their hard hats.
- Unsanitary toilet facilities.
- Turning off the lights at night.
- Failing to mail letters.

2.6.3 Risk review meetings

Risk review meetings should be organised regularly with decision stakeholders, the purpose being to:

- Discuss the results of regular risk inspections.
- Discuss the implementation of a decision with the aim of identifying potential new threats or opportunities.
- Maintain effective communications with decision stakeholders.
- Facilitate cooperation in instigating, developing and implementing measures to minimise threats and maximise opportunities.
- Formulate, review and disseminate standards, rules and procedures to ensure that the decision outcomes are achieved, ideally better than planned.

2.6.4 Industry information

Ensuring that decision-makers keep themselves up-to-date with the latest industry information relating to new research and practice is important in highlighting new threats and opportunities to decision outcomes. This can be done through regular training programmes, email lists, websites, risk and opportunity news letters, and bulletins etc.

2.6.5 Automatic sensors

It is important to use any technology available to monitor and detect potential physical risks that might arise in the workplace. Such risks include noise, dust, fire, fumes, vapours, gases, temperature, sun, radiation, security etc. Appropriate sensors should be installed which are connected to automatic controls or communication systems that can facilitate a response. Sensors might include, heat sensors, dosimeters, static area monitors etc. and control devices might include alarms, sprinklers, ventilation fans etc. It is important that such equipment is inspected, tested, recalibrated and maintained regularly.

2.6.6   Incident investigations

Incidents are defined as the occurrence of any event which causes actual loss or benefit to business objectives. Thorough investigations of any incidents after they have occurred are important since the lessons learnt can be very useful in preventing a repetition of events and in identifying further potential risks and opportunities to business activities.

2.6.7   Performance appraisals

Regular performance appraisals should be conducted using the objectives, KPIs and measurement criteria identified at the start of the risk identification process (see Section 2.4). Deviations from planned performance levels may indicate new risks or opportunities which will need investigation, analysis and response.

2.7   Conclusion

All decision-making is based on imperfect information regarding future uncertainties which can spoil the best intentions of decision-makers. The object of risk identification is to identify these uncertainties so that they can be managed. In essence, this involves three main steps:

1   Identifying and prioritising decision objectives.
2   Identifying assumptions made about the future in achieving those objectives.
3   Questioning those assumptions.

For many, the acceptance of risk identification processes will require a change of mindset from reactive to proactive management. Risk identification should be a continuous process which commences when a decision is being made or a project is being initiated and continues until a decision or project is fully implemented and its objectives achieved. The idea is to detect risks before they are created rather than to detect them after they have arisen. To be most effective, risk identification should be integrated into the culture of a project, into every corner of its decision-making processes, at all hierarchical levels and at all stages of its life cycle. This chapter ends with two case studies which illustrate the practical application of the techniques discussed above. One looks at the risk identification process on a joint venture project to construct a major new wastewater treatment plant in Australia. The other looks at how one company successfully managed a series of bomb threats on a very large commercial building project in Australia.

2.8   Case study – Risk in joint ventures

In this section we will analyse the process of risk identification on a major joint venture project to design and construct a new wastewater treatment plant in Australia. The strategy behind the development was to improve wastewater management in a region over the next 20 years. KPIs included improving ocean water quality by reducing wastewater discharges and increasing water reuse. This case study considers the issue of risk identification from the perspective of one joint venture company that was formed specifically to tender for the project.

2.8.1 Project history and scope

The need for a new wastewater treatment plant in the region had been recognised for some time. However, it was not until a large local steel company agreed to purchase water from such a plant, that the project became feasible. The contract for this project involved designing and constructing a new water treatment facility and laying a series of major pipelines from neighbouring towns and a new long-reach sea outfall. The facility was to be built in a densely populated area of Australia and, due to the major earthwork operations involved, had attracted significant public concern in its planning. Another problem was the close proximity of a prestigious golf course and areas of cultural and ecological significance along planned pipeline routes. An important requirement was that the tender should consider and guarantee the operational costs of the working plant and there were substantial compensatory arrangements for cost overruns should the contract be awarded and the plant be built. All risks except *force majeure* were to be borne by the contractor. The duration of the project from award to completion was 18 months and the tender period was 120 days.

2.8.2 Project stakeholders

The client was a Regional Water Corporation and the Joint Venture Company comprised of three partners in a 47.5 per cent, 47.5 per cent, 5 per cent equity-share arrangement. There was also one major sub-contractor involved in the concept development and tender process and the relationships of the parties are illustrated in Figures 2.7 and 2.8. The objectives of each stakeholder are summarised in Table 2.2.

2.8.3 Establishing a risk identification team

The first risk decision was whether to invest time and resources in tendering for the project and the Director of the relevant business unit made it on the advice of senior managers. The broad criteria considered in making this decision were profit potential, available resources and expertise, existing workloads, client importance and reliability and alignment with company strategic plan. The risks involved were not formally identified at this stage

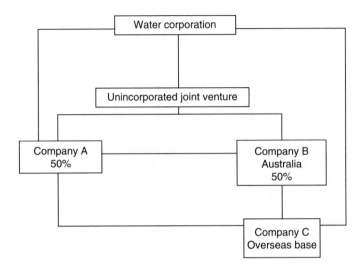

Figure 2.7 Parties to the project.

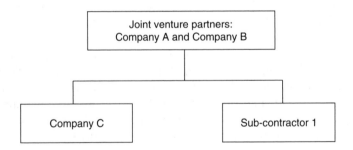

Figure 2.8 Joint venture risk distribution.

and the decision was based largely on the experience and judgement of senior managers – "the managers just knowing whether or not it was a good project to bid for".

Having made the decision to bid for the project, a tender team was established. This was done at two levels (organisational and individual) and on the basis of a broad review of project risks. The main risks were geological, environmental, community, legal, operational and technical.

2.8.4 Organisation

Company A was immediately interested in tendering for the project since the project was aligned with the business plan of one of its business units

Table 2.2 Stakeholder objectives

Stakeholder	Objectives	Associated issues
Client	Successful completion of the project at minimum costs. Deference of financial risks.	Desire to contract out all financial and political risk and for client to maintain good public relations.
Company A and B	Successful completion of the project at maximum profit. Minimisation of risks.	The project is extensive and highly sensitive. Could raise major PR issues and associated risks.
Company C	Successful completion of the project with zero risk.	Desire to work under a fixed-fee arrangement with no risk.
Sub-contractor	Successful completion of the project at maximum profit.	Potential conflict of interests with JV company.
Community	Successful completion of the project with minimal disturbances to environment, heritage and archaeological sites and community facilities.	Management of the community is important with sensitive issues involved.

and it had recently successfully completed a similar project. Considering the project's scale, complexity, high risks and specialist nature, Company A initiated a Joint Venture (JV) agreement with Company B which also had experience of operating in this area. Company C was also involved on the basis of being a well-known and respected engineering company that could provide the services of specialist engineers and a marketing edge through its excellent reputation and an established relationship with the client. It was involved as a minor JV partner with a no-risk contract linked to its perceived inability to influence project outcomes. Company C was happy with this arrangement, as were Companies A and B, as they did not want Company C to have any influence over project strategy.

Senior managers from Company A, B and C met to create a special arrangement to recruit, from existing projects, their "best" people to work on the tender. Availability was not an issue and team members were selected on the basis of technical expertise, prior experience and previous success in dealing with this type of project. The team operated under a distinct name to provide it with a strong self-identity and consisted of about 25 people which was split into smaller "work groups" of 3–5 people working on specific tasks. Specialists were consulted as the need arose. Once selected, team members were briefed on the project, their role and company objectives, vis-à-vis this particular project bid. This was deemed important to place the project risks in context. At this stage, the objectives of external stakeholders were formally considered through internal experts from Company A. Examples included environmental and community consultants who remained for the duration of the project.

2.8.5   The risk identification process

The bid preparation process was largely informal because the JV arrangement made the formal tendering systems developed in each partner company difficult to apply in a cross-company context. Company A's tender process was most influential although it contained no formal provisions for risk identification as a distinct process. Consequently, throughout the tender formulation period, risk identification was not a conscious or deliberate process and it was not underpinned and guided by formal procedures or techniques. Rather, the tender team met weekly in "project control meetings" to review the progress of the bid and it was then that risk identification was carried out. However, these meetings provided little scope for creativity and there was little sense of leadership relating to risk identification issues. Instead, the meetings were consumed with reviewing progress and factual issues relating to the exchange of information between team members. Without exception, risk identification was a deductive process based upon the experiences, abilities and subjective judgements of individual tender team members, the process being voluntary rather than mandatory. Inductive techniques such as brainstorming – for the sole purpose of identifying risks – were not used and, in the main, discussions were kept internal to the project team, external specialists only being consulted in exceptional circumstances. It is likely that this was a consequence of the time pressure that the team was under to produce the bid.

Although no formal guidance was given to the identification of risks during the bidding process, there was an expectation by senior managers that risks were being considered. To check this, a series of "risk review meetings" were organised "to ensure that every facet of the project bid was optimised". These meetings occurred towards the end of the tender process and focussed upon specific issues such as programme, delivery method etc. and were chaired by a Company Director from Company A, who had responsibility for evaluating risks for most of its projects. Typically, each risk review meeting was attended by between 10 and 20 people including senior bid team members, the General Manager of the business unit undertaking the work and the Chief Legal Counsel. The purpose was for each member of the bid team to formally present the major risks which they perceived to exist on the project and for these to be openly discussed by the General Manager, Chief Council and Director. It was also expected that the risks be presented to management with an appropriate mitigation strategy. Although debate occurred in these risk review meetings, it was seen as a bad sign suggesting that the bid team had not done their work properly. This made the risk review meetings an intimidating process for those making presentations and the meetings were largely driven by the agendas of the senior team members who were generally very busy and who did not want to waste their time in lengthy debates. During these meetings a "four element approach" to risk identification was adopted, which involved splitting the project into four stages (pre-tender, tender negotiations, construction, commissioning/operation), bid team members listing the major risks they perceived in each. The risks were listed with consequences and probabilities estimated, with a mitigation strategy outlined. Subsequent discussions revolved around these issues, during which there were no formal techniques used to elicit ideas, the process being *ad hoc* and driven by experience and seniority. This meant that further risks were unlikely to be identified during this review process or after the review, bid team members being fearful of being punished because they had not been identified earlier. In this sense, the risk review meetings performed an auditing rather than supporting function, meaning that new risks were often suppressed or "swept under the carpet".

2.8.6 Conclusion

A systematic and structured approach to risk identification was missing from this project despite being a major and complex undertaking with a high level of potential risk undertaken by a multinational JV company. All companies involved in this JV had a generic risk management system that placed little emphasis on risk identification as a separate and distinct element of the risk management process. This is typical of many companies in the construction industry. Consequently, risk identification was an incidental and fortuitous process of deductive reasoning where team members relied upon their own knowledge and experiences and upon the creative power of social interactions with other team members. Intuitive approaches to risk identification were restricted to information searches rather than to any specialised creativity techniques such as brainstorming or influence diagrams. In this informal environment, the structure of the bid team in terms of the dynamics it created was essential to the risk identification process. Since it is notoriously difficult to build effective teams, the risk identification process was left largely to chance, the JV companies being lucky with the depth of experience at their disposal. Smaller companies relying upon the same methods would have fewer resources to draw upon and be far more vulnerable to overlooking potential risks. While it is not desirable to sterilise the risk identification process by eliminating all subjectivity, it is likely that this project would have benefited from a more structured and inductive approach to risk identification, based upon the conscious and systematic application of established techniques.

2.9 Case study – Responding to a bomb threat

In this section we will analyse the process of risk identification, analysis and control on a large multistorey commercial project in the centre of Sydney. This case study considers this process from the perspective of the contractor's project manager who had received a number of bomb threats. The timing of this problem was particularly sensitive given the recent commencement of the war in Iraq and the bombing in Bali.

2.9.1 Background

This project was part of a very large development in the centre of Sydney which comprised three separate commercial buildings of over 50 storeys each. The contractor in this case study was responsible for one of the buildings. At the time of the bomb threats, an adjacent tower being built by another contractor was nearing completion. Construction of the third tower had not yet commenced. Three bomb threats had been received over a period of several weeks. Two of these threats had been non-specific and may have been directed at an adjacent project. However, one recent threat had been aimed directly at the contractor's project, specifying that the bomb was planted on a certain floor level, but providing no specific location. The relevant authorities and police have been informed on each occasion and the telephone calls had been traced to a specific telephone box in an adjacent street.

On each occasion, the bomb threats had resulted in a decision to evacuate and inspect the site. Apart from the obvious risks posed to employees and the public by the possibility of a bomb exploding, the process of evacuating over 400 workers presented its own significant risks to employee and public safety and to the security of plant and machinery on the site. There were also significant financial and programme implications each time the site had to be evacuated.

The problem facing the contractor was how to respond to these bomb threats in a rational way while protecting the safety of all site personnel and the contractor's commercial interests. To this end, the contractor's risk management system was used to identify clear objectives and develop a response plan to meet them.

2.9.2 Stakeholder analysis

The first step in the risk management process was to identify the variety of stakeholders who may have an interest in the outcome of the process and classify them. This would enable an appropriate stakeholder consultation plan to be developed for their involvement. The following stakeholders (Figure 2.9) were identified in a one-hour meeting between the project manager and senior managers of the contractor.

In accordance with the guidance provided in the company's risk management system, the stakeholder consultation plan involved:

- Key stakeholders – to be consulted individually by meeting, to identify their objectives in responding to this plan.
- Important stakeholders – to be kept informed of the ROMS process.
- Minor stakeholders – to be informed of the ROMS process if considered necessary.

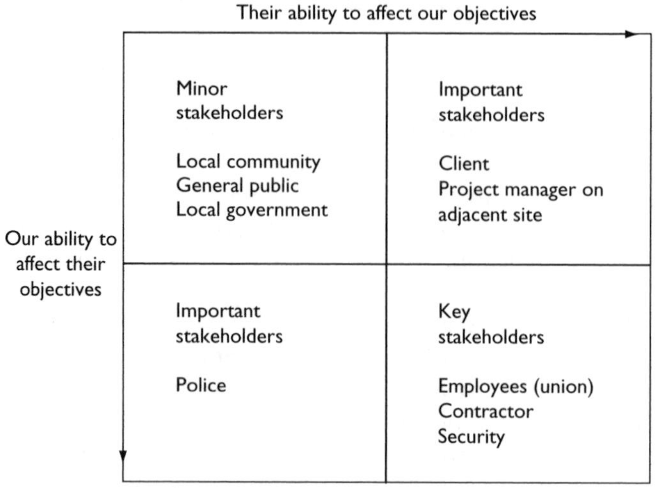

Their ability to affect our objectives

	Minor stakeholders Local community General public Local government	Important stakeholders Client Project manager on adjacent site
Our ability to affect their objectives	Important stakeholders Police	Key stakeholders Employees (union) Contractor Security

Figure 2.9 Stakeholder analysis.

2.9.3   Risk identification

In order to better understand the risks associated with the bomb threats, a one-day brainstorming workshop was arranged with key stakeholders. The purpose of this workshop was to identify and assess the risks and opportunities associated with the problem. An external facilitator was arranged and managed the process in consultation with the contractor's industrial relations manager and project manager. Before the meeting, stakeholders were asked to supply any relevant information. This was then distributed to all stakeholders for consideration and ensured that people had time to think about and comment on the problem and documentation before hand. The key stakeholders were also asked to list any risks and opportunities in advance and bring their ideas to the meeting for discussion.

The room used for the brainstorming workshop was a meeting room on site. This venue accommodated everyone comfortably around a single table which would help to encourage discussion and debate. An agenda was circulated in advance of the meeting. This reiterated the objective of the workshop and provided enough flexibility and time for everyone to contribute meaningfully. On the day of the brainstorming workshop, the room was set up to create a relaxed atmosphere which would engender open discussion. There were scheduled breaks for refreshments, food and refreshments were provided and the records of all deliberations and decisions were recorded on standard forms using flip charts and a PowerPoint presentation for all to agree and sanction. Mobile phones were switched off, and no disturbances to the meeting were permitted unless there was an urgent problem.

The meeting started with a professional facilitator explaining the purpose of the process and the role of each participant. There was also an introductory talk given by the project manager to remind everyone of the problem, any recent developments and his expectations of the process. Due to an ongoing positive relationship with the unions on this site, there was a constructive atmosphere and a desire to resolve the problem to everyone's satisfaction.

The brainstorming process commenced with each stakeholder identifying their objectives in resolving the problem. To focus minds, each stakeholder was restricted to five objectives. These were recorded on the PowerPoint presentation and then common objectives were listed. Once again, this list was restricted to a total of five common objectives. Later in the risk identification process each common objective would need to be broken down further. Therefore, more than five would make the process too lengthy.

Identifying common objectives was a challenging and revealing process. This is natural and to be expected, since different stakeholders are bound to have different expectations and perceptions of the risk management process. However, through discussion, it became evident that it was possible to identify five common objectives and agree a priority. These then became the focus for the rest of the risk identification process, creating a sense of collective responsibility for the resolution of the problem. Teamwork is critically important for the effective management of risks, and the importance of coming to an agreement about a common set of objectives at the start of the risk management process was recognised.

After there was agreement on the list of common objectives and their relative priorities, KPIs were identified for each objective. This helped to clarify the performance standards that were being sought and helped to further draw out subtle differences between the stakeholders' understanding of the problem and expectations of its solution. These KPIs then became the focus for the next stage of the brainstorming process which involved identifying the risks and opportunities associated with each KPI. They were recorded on the *Objective Identification Form* provided in Table 2.3 and signed-off in the meeting to ensure that everyone knew they were agreeing to a set of common objectives and KPIs. In all, the process of identifying and agreeing objectives and KPIs took about one hour.

The process of identifying risks and opportunities commenced with each stakeholder listing the risks and opportunities they had previously identified. These were recorded on a flip chart and then common risks and opportunities listed. There was then an open discussion, managed by the facilitator, when enabled further risks and opportunities to be identified. No restrictions were placed on ideas, and honesty and openness were encouraged. This was quite confronting for participants since they or their company could have been identified as the source of risk by other participants and even

Table 2.3 Objective identification form

Describe decision: How to respond to bomb threats in a rational way while protecting the safety of all site personnel and the commercial interests of xxxxx Construction Ltd.		Date: 02/09/2003	Ref: 2.5.1
List stakeholders	**List stakeholder objectives** (see pages 24–25)	**Define key terms**	**Rank common objectives**	**List KPIs** (see page 27)

List stakeholders	**List stakeholder objectives** (see pages 24–25)	**Define key terms**	**Rank common objectives**	**List KPIs** (see page 27)
Contractor	Reduce/eliminate bomb threats		3	Decrease hoax calls
	Maintain safety of site personnel in evacuation		1	Zero incidents from bomb hoaxes
	Minimise disruption to site activities in evacuation			
	Avoid having to automatically respond to every bomb threat		2	Acceptance of procedure/solution
	Make security everyone's responsibility		3	
Union	Reduce/eliminate bomb threats		1	Increased vigilance/number of security incidents
	Maintain safety of site personnel			
	Involve workforce in solution		2	Level of cooperation
	Make security everyone's responsibility		1	
Security company	Reduce/eliminate bomb threats		3	
	Maintain safety of site personnel in evacuation			
	Reduce breaches of security		2	
	Ensure efficient evacuation			
	Make security everyone's responsibility			
Compiled by:				Date: 02/09/2003
Reviewed by:				Date: 02/09/2003

themselves (if the process of facilitation worked effectively). Any potential risks and opportunities were recorded, however outlandish they seemed, and added to the list. This process took approximately two hours. After a 15 minute break, these ideas were then condensed into a shorter list which then became the focus for the next stage of the risk management process – risk analysis. They were recorded on the *Risk and Opportunity Identification Form* (as shown in Table 2.4) and signed-off in the meeting to ensure that everyone knew they were agreeing to a set of risks and opportunities which could then be analysed in more detail. Potential opportunities were recorded in italics. In all, the process of identifying and agreeing a list of risks and opportunities took about two hours.

2.9.4   Risk analysis

The risk analysis process involved systematically working through each of the risks and opportunities identified and recorded. Given the unusual and urgent nature of the problem, the lack of data available on similar problems and the "soft" nature of many of the potential risks and opportunities associated with it, a qualitative analysis was most appropriate. However, where reliable data was available and where it made sense to quantify the impact of a risk, numbers were also used in the analysis.

The process of risk analysis involved first considering the adequacy of existing controls for each risk and opportunity. This process alone was quite revealing and highlighted the strengths and weaknesses of current management systems. Then, taking these existing controls into account, the company's qualitative risk matrix was used to attribute a value to the risk or opportunity. This involved each participant estimating the probability and consequences of each possible event separately and then discussing them to reach an agreed rating. The imminence of each risk and opportunity was also discussed and identified to avoid the common and important pitfall of assuming that a low probability event cannot occur immediately. When this process had been completed, each risk and opportunity was ranked into priority order so that the project's resources could be deployed most effectively in dealing with this problem. Top priority was given to high and imminent opportunities and risks and lowest priority was given to low risks and opportunities which were not imminent. It is very interesting that this process resulted in an opportunity rather than a risk being top priority for the contractor. This information was recorded on the *Risk and Opportunity Register* (Table 2.5) and signed-off in the meeting to ensure that everyone knew they were agreeing to a prioritised list of risks and opportunities which could form the basis of the next part of the risk management process – risk control. Potential opportunities were recorded in italics. In all, the process of analysing the risks and opportunities associated with this problem took about two hours.

Table 2.4 Risk and opportunity identification form

Describe decision: How to respond to bomb threats in a rational way while protecting the safety of all site personnel and the commercial interests of xxxxx Construction Ltd.		Date: 02/09/2003	Ref: 2.5.1

List KPIs (From Appendix E)	List risks and opportunities to KPIs (see pages 28–47)	How could a risk or opportunity arise? (Describe causes and sequence of events) (see pages 28–47)
Zero incidents from bomb hoaxes	Bomb threats	Threats communicated to Police, project or neighbouring site. Currently, there are no patterns discernable in threats to indicate motive. Possible motives – employees wanting time off (since established/traditional response to bomb threat is to evacuate), theft during evacuation, local businesses may gain from evacuations, extortion/blackmail possible in the future to stop threats
	Sabotage (copycat threats)	Employees not being aware of serious punishment/implications, time pressure, tiredness, explode spray cans. set off ramset charges
	Open access to site – ability to create threats	Poor security (inadequate detection technology, too many entrances, poor entrance security checks, turnover in security staff, untrained security staff, disinterested security staff)
	Evacuation	Untidy site, unclear access, poor house keeping, poor lighting, poor signs, poor understanding of evacuation procedures
	Over reaction (panic)	Poor communication of hoax, rumours, poor/slow evacuation procedures, no grading of threats (all threats seen as serious)
	Under reaction (complacency)	Constant evacuation in response to every false hoax, unaware of seriousness of threat, no grading of threats
Acceptance of procedure/solution	Poor communication Perception of poor motives	Make decision in isolation, poor management practice, no clear authority General lack of investment in safety, time pressure, unsympathetic managers, leadership (managers not taking threats/procedures seriously)
	Interference with work	Bureaucratic response, constant evacuation
	Lack of trust/credibility	No consultation, perception that managers don't care, workers thinking cost has been prime factor in driving response, unrealistic expectations of managers (bureaucratic response)
	Rewards (monetary/ non-monetary)	Notice of potential external and internal rewards/benefits of reporting

Increased vigilance/ number of security incidents	Poor leadership No appreciation of risk attached to level of threat Poor upward communication No time	Managers not taking bomb hoaxes seriously, managers not listening to workers Poor education/training/communications, constant false hoaxes, poor response by authorities, lack of education, feelings of embarrassment in false alarms Poor communication systems for potential threats, no clear reporting responsibilities Bomb threats a low priority in relation to other goals/pressures
Decrease hoax calls	External appearance of panic *Identification of suspect*	Lack of coordinated evacuation *Police investigation, encourage workers to be vigilant and report potential threats, rewards for information*
Level of cooperation	Lack of trust in management	No consultation, perception that managers don't care, workers thinking cost has been prime factor in driving response, unrealistic expectations of managers (bureaucratic response)
Compiled by:		Date: 02/09/2003
Reviewed by:		Date: 02/09/2003

Table 2.5 Risk and opportunity register

Describe decision: How to respond to bomb threats in a rational way while protecting the safety of all site personnel and the commercial interests of xxxxx Construction Ltd.					Date: 02/09/2003	Ref: 2.5.1
List risks and opportunities (Copy from Appendix F)	Describe existing controls and rate adequacy Inadequate Adequate Good Excellent	Probability (Probability Label from Table 5, page 55)	Consequences (Consequence Label from Table 6, page 55)	Level (Overall level of risk from Table 7, page 56)	Imminence Occurred Imminent Not imminent	Rank risks and opportunities
Bomb threats	Inadequate	Possible	Major	High	Not imminent	2
Open access to site	Good	Possible	Major	High	Not imminent	3
Sabotage	Inadequate	Rare	Moderate	Low	Not imminent	14
Evacuation	Good (Formal plan, communicated in site induction)	Rare	Moderate	Low	Not imminent	15
Over reaction	Good (Good communications and trained staff)	Possible	Moderate	Medium	Not imminent	9
Under reaction	Good (Good communications and trained staff)	Rare	Moderate	Low	Not imminent	16
Poor communication	Adequate	Possible	Moderate	Medium	Not imminent	10
Perception of poor motives	Inadequate	Likely	Moderate	Medium	Not imminent	11
Interference with work	Good	Likely	Moderate	Medium	Not imminent	8
Lack of trust/ credibility	Adequate	Possible	Moderate	Medium	Not imminent	12
Rewards (monetary/ non-monetary)	Inadequate	Almost certain	Minor	Medium	Not imminent	7

Poor leadership	Excellent	Unlikely	Major	Medium	Not imminent	13
No appreciation of risk attached to level of threat	Inadequate	Likely	Major	High	Not imminent	4
Poor reporting/ inspection	Adequate	Possible	Extraordinary	High	Not imminent	5
External appearance of panic	Inadequate	Almost certain	Major	High	Not imminent	6
Identification of suspect	*Adequate*	*Possible*	*Major*	*High*	*Not imminent*	*1*

Compiled by: Date: 02/09/2003

Reviewed by: Date: 02/09/2003

Table 2.6 Risk and opportunity treatment plan

Describe decision: How to respond to bomb threats in a rational way while protecting the safety of all site personnel and the commercial interests of xxxxx Construction Ltd.				Date: 03/09/2003	Ref: 2.5.1

Risks and opportunities in priority order (from Appendix I)	Recommended controls in priority order (see pages 69–75)	Preferred controls	Risk/ opportunity level after control High Medium Low	Cost/ benefit analysis (insert $ and accept or reject)	Person responsible for implementation of chosen control (see page 76)	Timetable for implementation (complete Action Plan – Appendix L for each person)
Identification of suspect	1. Improve control of access and egress from site (e.g. identity cards, security checks, single point of entry etc.). 2. Offer rewards for information leading to successful prosecution. 3. Improve communications with neighbouring site.					
Bomb threats	4. Information campaign (by poster or otherwise) to communicate security upgrade, how seriously xxxxx, Unions and Police take the issue, negative implications for community and fellow workers (theft, disruption, jobs, stress etc), severe punishment for perpetrator, encouraging vigilance, possibly exaggerate security upgrade to scare perpetrator – consult Unions and Police. 5. Eliminate potential gains for perpetrator and dictate response rather than hoaxer (Develop graded response to avoid automatic evacuation, increased communication with Police and neighbouring site, establish policy in case of attempted extortion).					

Open access to site	6. Improve control of access and egress from site – See 1 above.			
No appreciation of risk attached to level of threat	7. Develop graded response to avoid automatic evacuation in consultation with Police and Unions (see 5 above). 8. Better input from Police/authorities on appropriate responses. 9. Work with Union delegates to develop best practices. 10. Education/induction/training of workforce re: real risks and graded response. 11. Communication campaign (by poster or otherwise) to inform of risks and graded response (see 4 above)			
Poor reporting/ inspection	12. Encourage culture of vigilance and collective responsibility for security and safety. 13. Make reporting responsibilities and lines of communication for potential security threats clear. 14. Refine existing emergency procedures and those for reporting potential security threats. 15. Poster campaign – see 4 above.			
External appearance of panic	16. Develop graded response policy to avoid automatic evacuation in consultation with Police and Unions (see 5 and 7 above). 17. Refine evacuation procedures to avoid external appearance of panic and chaos. 18. Communicate proactively with media in order to avoid public perception of major crisis. Do not hide information from media.			

Table 2.6 (Continued)

Rewards *(monetary/ non-monetary)*	19. Offer as monetary a reward for information leading to successful prosecution (see 2 above). 20. Information campaign (by poster or otherwise) to communicate negative implications of threats for community and fellow workers, severe implications for perpetrator (see 4 and 11 above). 21. Lead and industry campaign to improve site security. 22. Develop a reporting hotline.			
Interference with work	23. Develop graded response policy to avoid automatic evacuation in consultation with Police and Unions (see 5, 7 and 16 above). 24. Refine evacuation procedures to prevent thefts and damage to work in event of evacuation.			
Over reaction	24. Develop graded response policy to avoid automatic evacuation in consultation with Police and Unions (see 5, 7 and 16 above). 25. Refine current evacuation procedures to prevent possibility of panic in event of evacuation. In particular, ensure facilities for speedy evacuation of employees from high levels.			
Poor communication	26. Demonstrate xxxxx have developed any policies and conduct any review of procedures in close consultation with Unions. 27. Information campaign (by poster, training, noticeboards or otherwise) to inform workers of any changes in practices (see 4, 11 and 20 above).			

Lack of trust/ credibility	28.	Demonstrate xxxxx have developed any policies and conduct any review of procedures in close consultation with Unions (see 26 and 27 above).				
Poor leadership	29.	Continue positive and open relationships with Unions and employees.				
Sabotage	30.	Information campaign (by poster, training, noticeboards or otherwise) to inform workers of how seriously xxxxx and Police take the threats and communicate penalties (see 4, 11, 20, 27 and 29).				
Evacuation	31.	Develop graded response policy to avoid automatic evacuation in consultation with Police and Unions (see 5, 7, 16 and 24 above).				
	32.	Refine current evacuation procedures to prevent possibility of panic in event of evacuation. In particular, ensure facilities for speedy evacuation of employees from high levels (see 25 above).				
Under reaction	33.	Develop graded response policy to avoid automatic evacuation in consultation with Police and Unions (see 5, 7, 16, 24 and 31 above).				

Compiled by: Date: 03/09/2003

Reviewed by: Date: 03/09/2003

2.9.5 Risk control

The process of identifying controls to mitigate each risk and maximise each opportunity involved systematically working down the newly prioritised list of risks and opportunities. Participants were not restricted in the number of potential controls identified for each risk and opportunity, which were recorded onto the *Risk and Opportunity Treatment Plan* (Table 2.6). After estimating and comparing the costs and benefits of each possible control, taking into account the reduction of risk and increase in opportunity it potentially provided, a preferred list of controls was identified in ranked order. Having costed each control in the cost benefit analysis, it was then possible to calculate the total cost of dealing with the bomb threats and allocate resources accordingly. Importantly, responsibilities for implementing each control were allocated to specific individuals and a timetable for implementation agreed. For each person identified in the treatment plan, an individual action plan was developed. These were monitored in regular site meetings, where a specific risk management item was included on the agenda.

2.9.6 Conclusion

The risk management process described above took approximately eight hours and produced a comprehensive strategy to deal with a very serious problem. Of course the implementation of the strategy took longer and was an ongoing process. Nevertheless, the perpetrator of the bomb threats was eventually caught and turned out to be a disgruntled employee working on the project. As well as contributing to the resolution of this problem, the risk management also helped to cement industrial relations on the project by allowing the contractor to demonstrate its commitment to open communication and the involvement of project stakeholders such as the unions in the resolution of major problems which affect their interests.

At the end of the process, each of the forms discussed above and any data, records and assumptions made in developing the strategy were deposited in a project-specific Risk and Opportunity Management Plan (ROMP). This acted as a record of deliberations and decisions for this problem and all other significant problems on this project.

Chapter 3

Risk and opportunity analysis

3.1 Introduction

Having identified a potential risk or opportunity, the next step is to analyse it. Risk analysis is the process of evaluating identified risks and opportunities to discover their magnitude, whether they merit a response, and how responses should be prioritised in the light of limited resources. It is here that one will find most of the jargon and mathematical techniques, which have become associated, often negatively, with the subject of risk management.

Unfortunately, many managers seem to have developed a somewhat misguided impression that risk analysis always involves complex statistical procedures and computer programs. Furthermore, due to an increasing obsession with numbers and computer analyses as a means of legitimising and justifying managerial decisions, the quantitative techniques of risk analysis seem to have received an inordinate amount of attention, compared to the equally important techniques associated with other stages in the risk management process. This is unfortunate, and it is important to remember that risk analysis is just one part of the overall risk management process. Furthermore, it is possible to conduct a risk analysis competently, consistently and comprehensively without the use of mathematics. Indeed, most risk analyses are carried out in two stages:

- *Stage one* – A qualitative analysis of risks and opportunities using qualitative/descriptive scales such as high, medium and low.
- *Stage two* – A quantitative analysis of risks and opportunities using numerical estimates. This is normally conducted on risks and opportunities which emerge as particularly important from stage one and where reliable data for analysis is available.

This chapter sets out a clear framework for dealing with both the quantitative and qualitative aspects of risk analysis. It is concerned with the process and techniques of risk analysis and with individual and corporate attitudes towards risk. To start we shall focus on quantitative risk analysis, where numbers rather than words are used to describe different levels of risk.

3.2 Quantitative risk analysis

While many managers and organisations like to use numbers to justify their decisions, the rigorous use of quantitative techniques to assess project and business risks is very limited outside the major hazard industries such as oil, chemical, petrochemical and nuclear power. This reluctance may be attributed to a widespread belief that the complexity and cost of risk analysis increases as it becomes more quantitative. It may also be due to a lack of quantitative data available to input into such analyses. For example, the construction industry is notorious for its poor record keeping. However, as business corporate reporting requirements become more stringent and demanding, companies can no longer afford to ignore the need for quantitative risk assessment – where appropriate. In particular, given the increasing availability of very affordable, high-powered laptop computers and user-friendly spreadsheet softwares, there are fewer acceptable reasons why managers should ignore a quantitative approach to assessing their business risks and opportunities. Quantitative risk assessments can now be carried out very simply, quickly and economically and, where appropriately used, can have significant advantages over qualitative methods, which can be very restrictive in scope.

Quantitative risk analysis can be undertaken competently, consistently and comprehensively with the use of very little mathematics, statistics and computer programming. However, a basic understanding of the rudiments of probability adds considerably to the focus of the professional who needs to make decisions under conditions of uncertainty. To this end, the following sections introduce basic notions of probability in a simple and easy to understand way. We also demonstrate how a rudimentary knowledge of probability can be practically used to help managers make decisions in risky situations.

3.2.1 Probability

As we discussed in Chapter 1, a probability is a number between zero and one, both inclusive, which represents a judgement about the relative likelihood of some event. Zero implies the event is impossible and one implies that it is certain. Probabilities can be derived from one or a combination of three sources.

First it may be an *objective calculation* based on observed relative frequencies of past incidences of the same event; for example, the number of times when there has been more than two centimetres of rainfall in November in central Sydney. This type of judgement is only reliable when it applies to identical and repeatable events in a strictly controlled situation where all variables can be perfectly controlled. Although frequencies are commonly used to calculate probabilities in project environments, the differences between project environments ensure that resulting objective calculations of risk and opportunity have to be treated with extreme caution.

The second source for assessing probability is where there may be some *a priori* basis, derived from some visible symmetry, for a particular event. This renders unnecessary the collection of frequency observations. For example, when one tosses a coin, the probability of flipping a head is 0.5 (assuming it is a perfectly balanced coin and ignoring the very small chance that it could land on its edge). Similarly, for a perfectly balanced die, the probability of throwing any number between one and six is 1/6 (0.17).

The third possibility of deriving probability is the so-called personalistic view or *subjective probability*. This is a reflection of consistent perceptions, opinions and judgements about an event. For example, people frequently make verbal statements about probabilities and choices about gambles, reflecting their own belief in the likelihood of an event. In the absence of an *a priori* basis for judgement and reliable frequency data based on previous identical and repeatable events, personal subjective probabilities allow just as meaningful a discussion of unique events as objective calculations do of repeatable events.

3.2.1.1   Personal subjective probabilities

We have argued that in most real world situations like a construction project, there is no possibility of repeatable events under substantially identical and controllable conditions. Similarly there is little possibility of gathering large sets of relative frequency data. Thus for most practical purposes subjective probabilities offer the only meaningful way of dealing with risks and opportunities encountered in project decisions.

There are two approaches to eliciting subjective probabilities, namely the *direct* and the *indirect* methods. The direct method assumes the existence of a rational decision-maker well aware of the rudiments of probability. It involves asking the subject to assign a number to his or her opinion about the likelihood of the outcome in question. In contrast, the indirect method involves asking a series of questions, from the answers to which it is possible to impute a personal probability. The most appropriate method will vary depending on the subjects being questioned and their knowledge of probability. For example, if you are a supervisor dealing with operatives who are likely to have little or no knowledge of probability you would use the indirect method, but if you are a manager dealing with professionals who are likely to understand probability, you would use the direct method.

Whatever method is used, someone ultimately has to make a judgement about probability levels and in doing this, it is worthwhile employing a range of practical strategies to maximise reliability and validity. In the interests of succinct advice, we shall need to make some fairly general statements. For example, managers have a tendency to overestimate low probabilities. Most professionals would prefer to warn of a suspected event and be proved wrong than not to have warned of it. Hence, when eliciting subjective probabilities it is good practice never to ask about the probability of a specific event without asking about the probability of its causal events, plotting these onto an "event tree". By doing this, one will invariably find that the joint probability of causal events will not add up to the judged probability of the end event (see Section 3.2.1.2 on joint and compound probabilities). It is also good practice to ask the same question a number of different ways in a search for inconsistencies in answers. If you find inconsistencies, be happy because they can be used as the basis for discussion with the respondent and they can reveal more detail about the risk or opportunity in question. Group-based questioning can often be useful in eliciting such inconsistencies. However, although there is a natural tendency to think that it is always preferable to use a group of experts to lay people, research indicates that inconsistencies may be greater with experts. Experts are often overconfident in their judgements and their self-knowledge (meta cognition) is often poor, compared to lay people. Furthermore, there is evidence to suggest that personal interaction between experts should be avoided. For this reason, the use of the so-called Delphi methods has been suggested, which enables a group of experts to have access to each others ideas without personal interaction. This helps to avoid potential bias derived from the overconfidence of particular individuals and from the effects of personalities, as opposed to technical skills, on the work of the group. Thus while eliciting a personal probability for a specific event, it is always worth considering why that judgement may turn out to be wrong and to create an environment where people feel comfortable in saying what they really think rather than what they should think. It follows that creating an open, trusting and no-blame organisational culture is as important as the above techniques in deriving accurate estimates of risk and opportunity.

3.2.1.2 Calculating probabilities

Calculating the probability of an event occurring is quite a straightforward process that involves an understanding of two simple principles, namely *joint* and *compound* probabilities.

JOINT PROBABILITY

The probability of a number of *mutually exclusive events* (independent events that cannot occur together) occurring is calculated by adding together the probabilities of the individual events. This is known as the joint probability of the set of outcomes. For example, if I place bets on two horses *in the same six-horse race*, then the probability of my winning is calculated by adding together the probabilities of the winning of both individual horses. If we unrealistically assume that each horse in the race has an equal chance of winning, then my probability of winning would be given by:

$$1/6 + 1/6 = 0.34$$

Logically, if you bet on every horse in the race then you are guaranteed to win (ignoring the small chance that none finish) and the probability is given by:

$$1/6 + 1/6 + 1/6 + 1/6 + 1/6 + 1/6 = 1.00$$

COMPOUND PROBABILITY

In contrast to joint probability which is relevant for calculating the probability of winning on horses *in the same race*, compound probability is relevant when betting *across different races*, where one outcome is *dependent* on the other. For example, if I placed a so-called accumulator bet, where I put money on horses in two separate races such that if the first horse wins then the winnings are placed on the horse in the second race, then the probability of my winning would be the product of the probabilities of each horse:

$$1/6 \times 1/6 = 0.028$$

This is a very much lower probability and a clear deterrent for placing accumulator bets. The probability is lower because winning on the second horse is *dependent* on winning on the first horse. This is a compound probability where we are interested in calculating the overall likelihood of both (interdependent) events occurring.

Compound probability has a well-known application in the construction field. For example, consider the budget forecast for a new library to service a rapidly expanding neighbourhood. This is summarised in Table 3.1.

Table 3.1 Budget forecast for new library

Budget component	Most likely ($)	Worst case (P=0.1) ($)
Site purchase	500,000	600,000
Substructure	250,000	310,000
Superstructure	750,000	950,000
Sub-total	1,500,000	1,860,000
Escalation	150,000 (10%)	279,000 (15%)
Totals	1,650,000	2,139,000

In Table 3.1, the budget has been broken down into four major headings. Each is given a best case estimate and a worst case estimate. The decision rule for deriving the worst-case estimate is that it should represent an event that should not occur more than once in ten times. Let us therefore assign a notional probability of 0.1 for each worst case. Assuming that the figures have been correctly estimated and that the components of the budget are independent of each other, then the probability of the project costing $2,139,000 is given as follows:

$$0.1 \times 0.1 \times 0.1 \times 0.1 = 0.0001$$

(site purchase increase $\times$ substructure increase $\times$ superstructure increase $\times$ escalation increase)

While the example is simplistic, it does illustrate that in considering the probability of a number of worst cases happening together, the odds rapidly diminish to very small numbers indeed – in this case it is 10,000 to 1. This simple principle is important and needs to be remembered for later use in this book.

3.2.2   Expected monetary value

It is often very useful for companies, in making decisions, to express their risks in dollars. Although this is not always possible with reliable accuracy (see Chapter 1) the resultant value is commonly referred to as the expected monetary value (EMV) of a decision.

When calculating EMV, it is important to appreciate that every event has a range of possible outcomes (consequences), each with a different probability of occurring. So far we have simplistically assumed that any event has only one possible outcome with an associated probability of occurring. This range of possible outcomes is called a *probability distribution*. For example, consider a lottery ticket which gives the owner a 0.75 chance of winning $5000 and a 0.25 chance of winning nothing. The expected monetary value (EMV) of the ticket is given as:

$$EMV = 0.75 \times \$5000 + 0.25 \times \$0$$
$$= \$3750$$

This implies that over a large number of transactions, I can *expect* to make $3750 from purchasing such lottery tickets. The significance of this EMV calculation is that it tells us there is no risk in spending $3750. It also tells us that if I spend more than $3750 then I can *expect* to lose money and the more I spend over this amount, the more risk I incur.

While valuable, it is important to appreciate that EMVs, when based on objectively derived probabilities, are only meaningful in the context of a large number of identical transactions. Unfortunately, it is sometimes used inappropriately to assess decisions of a more unique nature, which change over time.

3.2.3   Risk attitude

We demonstrated in the previous section how, with a rudimentary knowledge of probability, it is possible to calculate the EMV for a decision. Logically, one would assume that decision-makers would wish to pursue the maximisation of EMV as a decision criterion when dealing with decisions under risk. However, it is frequently observed in practice that people will sometimes prefer an alternative to the decision that offers the highest EMV. Utility theory offers a useful model for understanding this behaviour.

Utility theory is based on the assumption that personal attitudes to risk and opportunity when making a decision are a reflection of individual trade-offs between gambles and certain pay-offs. This trade-off is commonly known as the *risk to reward ratio*. In other words, where the risks of an investment are high, the investor will expect an additional premium to make it worthwhile carrying such a risk. There are many examples of this principle in operation. One of the most commonly quoted is the difference between the relatively low rates-of-return from relatively risk-free, government short-term Treasury Bills compared to relatively high rates-of-return from ordinary shares on the more volatile stock market. Another familiar example is motorcar insurance premiums. If you are a young person driving a fast car, then you can expect your insurance premium to be relatively high to reflect the greater risk to the insurance company for covering you. This principle also drives the pricing of investment loans in financial markets. For example, if an investment project is risky then the bank or finance house will add a risk premium to the base interest rate. The theory is that over time, the higher rate on the successful but risky loans will compensate for those few loans which become bad debts to the lender.

In terms of risk to reward ratios, individuals may be thought of as falling into three categories: *risk averse, risk seeking* and *risk neutral*. The differences between these categories can be explained by reference to the so-called basic reference lottery ticket. For example, suppose you were lucky enough to hold in your hand a lottery ticket which gave you an even chance of winning $10,000 or nothing at all. What is the lowest price you would accept for it? The answer to this question when compared to the EMV of the ticket reveals much about your risk attitude.

The EMV of this ticket is given by:

$$EMV = (\$10,000 \times 0.5) + (\$0 \times 0.5)$$
$$= \$5000$$

This means that you would be a *risk neutral* person if you were willing to sell the ticket for a minimum price of $5000. In contrast, you would be a *risk seeking* individual if you would not be willing to part with the ticket until someone offered to pay you well over its EMV, say $6500. In this case, $1500 ($6500–$5000) reflects the utility (value) you place on the thrill of the 50 per cent gamble that by holding onto the ticket you might win $10,000. Conversely, you would be *risk averse* if you would sell the ticket at a value less than $5000, say $3500. In this case, the $1500 reflects what you are prepared to pay (forgo) to avoid worrying or fearing about the 50 per cent gamble that by holding onto the ticket you could win nothing. This reflects a natural fear of uncertainty and/or a pessimism that things are more likely to go wrong than right (a focus on the downside of risk). In contrast, a risk seeking decision would reflect a high tolerance of uncertainty and an optimistic mindset which focusses on the upside of risk. Essentially, a risk seeker believes that if you are not in the game you cannot win but the risk averse person believes that if you are not in the game you cannot lose. These are fundamental differences in states of mind which can be intuitive but which can also arise from controllable factors such as skill levels in managing risks or organisational cultures that encourage a certain attitude towards risk. They are graphically illustrated in Figure 3.1 which shows how a risk averse person gets the same utility (value) from a low price as

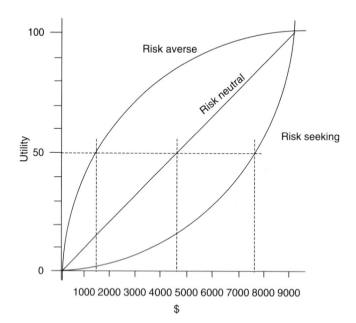

Figure 3.1 Utility curves.

the risk taker gets from a higher price. It is important to recognise that one individual may exhibit all three characteristics at different times, for different types of risk decision or for decisions involving different monetary amounts.

Utility theory explains how, but not why, rational people sometimes prefer outcomes that do not have the highest monetary value. It suggests that instead of maximising EMV, people maximise their own utility. The equation that describes the utility curve is called the utility function and this varies from person to person. In theory it is possible to plot a person's utility curve and to derive their utility function, but in reality it is extremely cumbersome and impractical – not least because it changes over time and depends on context. For this reason, a general awareness of this concept and informal allowances are preferable. To this end, let us explore the practical application of utility theory a little further with an example from construction.

Suppose we are currently negotiating two projects on behalf of a firm of contractors. The firm is currently working near capacity so we chose only one of the two projects. Project 1 is a low risk venture offering an almost certain chance of $500,000 profit. Project 2 is rather risky in that it involves taking design responsibility for some complex tidal protection works. Our calculations show that on Project 2 we have an 80 per cent chance of generating $1,000,000 profit and a 20 per cent chance of a $300,000 loss.

The EMV for Project 1 is $500,000 and the EMV for Project 2 is $740,000 $[(0.8 \times 1,000,000) + (0.2 \times -300,000)]$. In strictly rational terms, the best option is Project 2 because it has a higher EMV. However, this is a rather simplistic analysis that depends on the general situation of the company and how much managers in that company value the certainty of outcome. For example, if the possible (although relatively small) loss of $300,000, would send the firm into bankruptcy, then the company might choose Project 1, even though the financial analysis tells it that the EMV is lower. In other words, the organisation's circumstances have made its managers behave in a risk averse manner. Therefore, we can see that a simple EMV calculation without a consideration of risk attitude may point the decision-maker in entirely the wrong direction.

3.2.3.1   The Allais paradox

The fact that people often choose apparently sub-optimal decisions can be explained further by reference to the so-called Allais paradox. This is named after the French economist Maurice Allais who is credited with many advances in the general theory of random choice and risk psychology.

It can be explained with a decision-maker who is faced with two questions as follows:

1 Do you prefer situation A or situation B?

   Situation A: Receive a certain $1 million
   Situation B: Receive a lottery ticket with a
      10 per cent chance of winning $5 million
      89 per cent chance of winning $1 million
      1 per cent chance of winning nothing

2 Do you prefer situation C or situation D?

   Situation C: Receive a lottery ticket with a
      10 per cent chance of winning $5 million
      90 per cent chance of winning nothing
   Situation D: Receive a lottery ticket with a
      11 per cent chance of winning $1 million
      89 per cent chance of winning nothing

These choices may be presented as a decision tree (Figure 3.2).

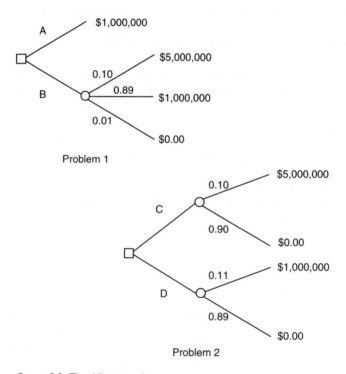

Figure 3.2 The Allais paradox.

It has been shown that over a large number of studies, people who are aware of probability calculus, who are considered to be rational and whose personal capital is small compared to the sums involved, have a natural propensity to prefer A over B and C over D. Yet the EMV of each choice is as follows:

EMV(A) = $1 million
EMV(B) = $1.39 million
EMV(C) = $0.50 million
EMV(D) = $0.11 million

This demonstrates how people can make sub-optimal decisions based on emotion rather than rational calculation. It also shows that people can unwittingly be risk seeking (in Question 2) and risk averse (in Question 1), depending on the context. In particular, people tend to make more sub-optimal decisions and become more risk averse when the figures involved are relatively large in relation to the subject's own capital and when there is one option with a high certainty of outcome. We shall return to the important subject of psychology in risk decisions, shortly.

3.2.4 Measuring risk exposure – Quantitative risk analysis techniques

The choice of technique for analysing risk depends on the size, type and the general nature of the problem being modelled, the amount and reliability of information available and the nature of the output required. In this section, we present eight different ways of dealing with risk analysis, which are appropriate in different contexts. The approaches are:

- The risk premium
- Sensitivity testing
- Expected monetary value (EMV)
- Expected net present value (ENPV)
- EMV using a Delphi peer group
- Risk-adjusted discount rate (RADR)
- Detailed analysis and simulation
- Stochastic dominance.

3.2.4.1 The risk premium

The risk premium is a rather coarse, but widely used, instrument which is also known as the contingency fund. Indeed, in industries such as construction, it would be regarded as negligent if any consultant produced an estimate or project forecast which did not include a contingency fund. This is testimony to the fact that in many industries such as construction, risks have long been accounted for as a matter of course. In these industries, the usual practice is to add a contingency premium to the base estimate to account for downside risks, accepted by the organisation which cannot accurately be forecast at the time. However, in practice, the way in which contingency allowances are calculated is often problematic. Calculations rarely take account of risk attitude, are often arbitrary and are not tailored to the specific project. For example, Yeo (1990) found that in the construction industry, many contingency estimates are seen as a routine administrative procedure underpinned by little investigation or rigour on the part of the estimator. Not only does this result in highly subjective estimates, there is a tendency to double-count risk because some estimators also subconsciously include for them in their base estimates. The result of these deficiencies is all too often the rejection of projects that are economic and the submission of overly conservative bids which are unsuccessful or inflated prices for clients when they are successful. A potentially greater problem is that such allowances can hide poor management and the potential for greater efficiency. So in summary, the risk premium is at best a rather blunt tool that is made less effective because it is also not used very effectively in practice.

3.2.4.2   Sensitivity testing

Sensitivity tests measure the effect on a decision output, of certain specified changes in the value of input variables (risks). For example, if the decision is to arrive at a contingency allowance for a tender, one may alter interest rates, energy costs, labour costs, construction period etc. as input variables to see what impact various percentage changes in each of these variables would have on project costs. This will reveal what input variables (risks or opportunities) project cost is most sensitive to. For example, a 5 per cent change in one variable may produce a 50 per cent increase in costs whereas a 5 per cent change in another variable might produce no change in costs. Clearly, the bigger risk variable which merits special attention is the one which produces the 50 per cent change. Furthermore, if costs increase when a variable in changed then it is a risk, but if costs fall then the variable is an opportunity (assuming that the objective is to minimise costs).

The resultant changes in a decision output (say, cost) arising from changes to input variables may be presented in tables, tornado, graphs or the so-called spider diagrams and allow the decision-maker to allocate limited resources appropriately to control the wide array of risks they may face in making a decision. Clearly, it makes no economic sense to allocate the same resources to the management of all input variables. It is a logical and efficient practice to ensure that those which have the largest potential impact should command the attention of the largest resources.

Normally, variables in a sensitivity analysis are considered to be independent and are altered one at a time. However, sensitivity analysis also allows the decision-maker to alter input variables in different combinations to see what impact there is on the decision output. Nevertheless, caution must be exercised in interpreting the results of such analysis since the likelihood of certain combinations can be very low. Now is the time to remember compound probability (see pp. 90). For example, if the decision model has three components and the probability of the worst case in each is, say, 0.1, then the probability of all three occurring simultaneously is:

$$0.1 \times 0.1 \times 0.1 = 0.001$$

Nevertheless, sensitivity testing, when interpreted correctly and conducted realistically, can convey an extremely useful picture of a project/investment decision under dynamic real-world conditions. Consider the following example which is an appraisal of a proposed rehabilitation and redevelopment scheme in America. The project involved the rehabilitation of an eightieth century church crypt to provide a museum and restaurant and the development of the surrounding churchyard to include speculative office space and a small (fee paying) nursery school. The sensitivity analysis consisted of three tables as shown in Tables 3.2, 3.3 and 3.4.

Table 3.2 Sensitivity test

Variable	Change in variable to eliminate profit (%)	New value of variable (approx.)	Original value of variable
Capital value of offices per square foot	25	$169	$255
Building costs (including ancillary costs p.s.f.)	37	$122 (office cost)	$89 (office cost)
Finance rate p.a.	302	74.4% p.a.	18.5% p.a.
Building period	350	4.5 years	1 year
Period before offices sold	700	2 years	0.25 year

Table 3.3 Sensitivity test scenarios

Variable	Original scenario	Optimistic	Realistic	Pessimistic
Growth in capital value	0	7%	5%	1%
Building cost increase	0	6% p.a.	7.5% p.a.	10% p.a.
Finance rate	18.5%	16.5%	18.5%	20%
Building period	12 months	9 months	12 months	15 months
Building before sale of offices	3 months	No delay	3 months	9 months
Contingencies	5%	5%	5%	5%

Table 3.4 Sensitivity test: changes in model output

Outcome	Original scenario	Optimistic	Realistic	Pessimistic
Residual profit	$749,540	$879,460	$779,650	$222,600
% of capital value	24.73	27.12	24.5	7.2
% of total costs	32.86	37.21	32.45	7.8
Increase over original estimate (%)	0	17.33	4.02	−70.3

Table 3.2 indicates that no profit would be made at all if, other things being equal, the capital value of the offices were to fall by 25 per cent on the original assumption. Similarly, the building period would have to increase from one year to two-and-a-half years, or the building costs increase by 37 per cent, from $89 to $122, for the profit to be entirely eroded. The table considers one variable at a time.

Tables 3.3 and 3.4 consider three scenarios with sensible changes in combinations of input variables. Table 3.3 shows the *ex ante* assumptions about a range of scenarios by taking, in each case, a set of sensible assumed values for each of the important decision variables. The pessimistic scenario assumes a period of high inflation and high interest rates and an oversupply of space with an unoccupied period of nine months before sale, a recessionary market view of the property market. The optimistic view is one of strong growth in the capital value of the finished project, with low inflation and interest rates. However, these are by no means the best and the worst cases, they are intended to be realistic combinations of variables in the market conditions prevailing at the time. Table 3.4 indicates that the original scenario for this project was risk averse in its assumptions about capital growth on the completed project. On building costs, it was slightly risk seeking since no contingency had been provided for cost increases in the original model.

There are several advantages to the use of sensitivity testing. It is quick and easy to use. It requires little information and it can usually be carried out by hand. Furthermore, it fully recognises uncertainty in the input variables and can show how the output will be influenced by changes in input variables either singly or in combination. However, there are also several limitations with this method. For example, it takes no account of the *likelihood* of the range of input and output variables. Therefore, it does not give a probabilistic picture of risk exposure and there is no explicit method of allowing for risk attitude. For this reason, it has been argued that the results of sensitivity tests are at best ambiguous and at worst mis-leading. They are said to be ambiguous because they do not suggest how likely it is that the pessimistic or optimistic results will occur. They can also be misleading when some analysts unrealistically take a number of very low probability worst or best case values of input variables and calculate the effect on the output. Such combinations produce extremely low probabilities, are very unlikely in the real world and such a test would produce exaggerated results. In response to these criticisms, Flanagan *et al.* (1987) developed a sophisticated method for integrating sensitivity and probability analysis in the solution of building life cycle cost decisions. However, while valuable, its complexity combined with the immaturity of risk management within industries such as construction has ensured that its use in practice has been extremely restricted. Therefore, for the time being, we believe that the use of non-probabilistic sensitivity testing, despite its limitations, would represent a major step forward in risks analysis capability for most companies.

3.2.4.3   Expected monetary value (EMV)

A simple way of incorporating probability into risk analysis is the EMV method. We have already mentioned this in Sections 3.2.2 and 3.2.3 when explaining risk attitude. Now we can take the concept further and demonstrate some practical applications. In this approach the decision is broken down into a number of components and each one is examined separately in terms of its probability distribution (a range of possible outcome scenarios with their associated probabilities). The probability distributions associated with each decision component may be as simple as two scenarios (best case and worst case) or it may contain a large number of scenarios ranging from pessimistic to optimistic. For example, consider the case of forward planning for a district health authority. In order for the health authority to make a bid for central funds, a budget estimate is required for a new clinic (neighbourhood is known – not actual site). The budget for bidding purposes might be broken down into only three components: substructure (affected by eventual ground conditions), superstructure (affected by weather) and inflation (affected by prevailing economic/market conditions). Estimators are then asked to think of three scenarios (best, most likely and worst case) for each tender component, attributing a probability and value to each scenario. It is also good practice to describe each assumed scenario in detail so that future EMV calculations can be adjusted for any changes in predicted scenarios. The only condition is that the combined probabilities of all three scenarios for any one component must add up to 100 per cent since they are mutually exclusive scenarios which cannot occur together. The outcome of this process is depicted in Table 3.5.

An optimistic (risk taking) bid, assuming all the optimistic scenarios happened, would be: ($720,000+$1,800,000)×1.06=$2,671,200. The probability of this happening would be 0.20×0.20×0.20=0.008 (0.8 per cent – very low).

A most likely (safe) bid, assuming all the most likely scenarios happened would be: ($800,000+$2,000,000)×1.07=$2,996,000. The probability of this happening would be 0.50×0.60×0.50=0.15 (15 per cent – moderate).

Table 3.5 Probabilistic scenario analysis of EMVs

		Optimistic	Most likely	Pessimistic
Substructure	Scenario	No rock found	Boulders found	Solid rock found
	Outcome	$720,000	$800,000	$1,000,000
	Probability	0.2	0.5	0.3
Superstructure	Scenario	Does not rain for whole contract	Rains for 15 days	Rains for 50 days
	Outcome	$1,800,000	$2,000,000	$2,500,000
	Probability	0.2	0.6	0.2
Inflation	Scenario	Reserve bank keeps rates on hold	Economy grows and Reserve Bank increases by one base point	Economy booms and Reserve Bank increases by two base points
	Outcome	6%	7%	8%
	Probability	0.2	0.5	0.3

A pessimistic (risk averse) bid, assuming all the pessimistic scenarios happened would be: $(\$1,000,000 + \$2,500,000) \times 1.08 = \$3,780,000$. The probability of this happening would be $0.30 \times 0.20 \times 0.30 = 0.018$ (1.80 per cent very high).

But pure scenarios are unlikely to happen and we cannot ignore the fact that every bid component has the possibility of a range of three values. To take this all into account it is possible to conduct an EMV calculation which involves calculating an average value for each component and combining the results. This is illustrated in Figure 3.3.

The above example is very simplistic. We have only identified two building components and the scenarios that could affect outcomes focus on one factor only – ground conditions in the case of substructure and weather in the case of superstructure. Of course, ground conditions and weather are not the only factors that can affect outcomes. Therefore, in reality the breakdown of an estimate would be far more detailed as would the range of possible conditions which could result in a particular scenario for each component (best, most likely or worst case). The level of detail used is the choice of the individual decision-maker and it is important to appreciate

EMV (sub) = $(0.2 \times \$720\,k) + (0.5 \times \$800\,k) + (0.3 \times \$1000\,k) = \$856,070$
EMV (sub) = $(0.2 \times \$1,800\,k) + (0.6 \times \$2,000\,k) + (0.2 \times \$2,500\,k) = \$2,060,000$
EMV (inf) = $(0.2 \times 6\%) + (0.5 \times 7\%) + (0.3 \times 8\%) = 8.90\%$

EMV (project) = [EMV (sub) + EMV (sub)] × EMV (inf)
$\qquad$ = [856,070 + 2,060,000] × 1.089
$\qquad$ = $3,175,600

Figure 3.3 The EMV for the hospital project.

that the description of scenarios and the attribution of probabilities and values is likely to be a highly subjective process – unless of course good quality data is available.

The advantage of the EMV method is that it considers all possible outcomes and avoids simply combining all the best and the worst cases to produce unrealistic extremes of possible outcomes (remember compound probability). The latter approach will merely produce answers that are too pessimistic on the upper bound and too optimistic on the lower. The EMV method is also suitable for a range of applications – budget figures, tender price forecasts, rates of project return or completion dates. It also overcomes some of the limitations of sensitivity testing by explicitly allowing for the probability of change in input values – producing a risk-adjusted outcome. The limitation of EMV, when based on objective probabilities, is that it is best used consistently over many similar-sized projects. The guidance it provides is helpful, but strictly, only in the very long-run.

3.2.4.4 EMV using a Delphi peer group

One of the issues in using any probabilistic technique is how to arrive at the probability values. We have already mentioned the value of the Delphi technique and here we explain how it can be used to assist in an EMV calculation. The Delphi method is named after the oracle at Delphi in ancient Greece. It utilises a formal Delphi group and is designed to pool the expertise of many professionals in order to gain access to their knowledge and technical skills while removing the influences of seniority, hierarchies and personalities on the derived forecast. It also eradicates the biases of poor meta-cognition and overconfidence which may encroach on expert forecasts. First, a group of experts is identified. The group members are kept separate to prevent any personal interaction, and the coordinator asks each member to make a forecast and a subjective probability estimate for the relevant components of the project or decision under consideration. The coordinator receives and summarises these estimates and the summary is given back to the members without any names attached. The group members are then asked to amend their forecasts in the light of the summary information. The new forecasts are then revised and communicated to all members. This process of forecast, summary, amendment and feedback continues until there is a consensus or when the members no longer wish to amend their forecasts. The result is the Delphi forecast and there is no doubt that this is a powerful method of assessing important projects at the budget and feasibility stage. In many projects it can easily be conducted using email over the course of one afternoon. The outcome would be something like the tables in the ENPV example below (see Tables 3.6 and 3.7).

The advantage of adding the Delphi group to the EMV method is that it is a well-recognised method of getting the best out of a group of experts in

Table 3.6 Net income projections

State of the house-building industry

Year	Declines ($)	Remains steady ($)	Growth accelerates ($)
Net income			
Year 1	170,000	180,000	200,000
Year 2	150,000	200,000	250,000
Year 3	150,000	200,000	250,000
Year 4	150,000	200,000	250,000

Table 3.7 Scenarios, probabilities and NPVs

State	Probability	NPV ($)
Steady state	0.6	142,680
Growth	0.2	280,200
Decline	0.2	14,420

a forecasting situation. The limitation is the extra resources and time it takes to undertake. Also, participants may not have a similar window of time in order to undertake the process simultaneously. Therefore risks and size of a project should be sufficient to warrant the effort required. Although it is worthwhile remembering that in the modern day version, there is no need for a two-day hike up a hot dusty mountain and no pigeon needs to die.

3.2.4.5   Expected net present value (ENPV)

The ENPV approach is useful in investment and development appraisal and can be applied in a wide range of situations where future income or cost flows are known. For example, it is used by the Victorian Government in Australia to evaluate tenders for PPP projects covering periods of up to 30 years. ENPV is also the basis of life cycle costing technique. ENPV is based on the combination of probability analysis and the corporate financial technique of discounted cash flows (DCF) which has been developed to convert future income or cost flows back to net present day values. The DCF technique is based on the assumption that the value of money diminishes over time due to a number of factors including inflation, taxation and earning potential if invested in the interim. These factors mean that a dollar today is worth more than a dollar in the future. This is reflected in ENPV calculations by using a *discount rate* (a percentage figure which reflects these factors) to convert future cost or revenue streams back to current day (present) values, thereby facilitating single point comparisons between different investment opportunities or risks. Essentially, an ENPV figure is the amount that would be needed today to purchase an equivalent amount of goods and/or services at some point in the future. So if a building component costs $1000 to repair in 10 years time, the ENPV of that repair cost is the equivalent amount it costs *today* to carry out that exact repair. Given that inflation will invariably increase repair costs over the 10 year period, the ENPV figure of an equivalent repair today will always be less than $1000.

So the discount rate can be based on a number of factors which determine how the value of money changes over time. These include future rates of inflation, taxation rates, affordability rates and investment rates (interest rates, bond rates or equity rates) that determine how a dollar invested now can grow in value over the period being considered. For example, the discount rate used by the UK government and Australian Victorian State Government for the economic appraisal of PFI and PPP projects respectively has been 6 per cent per annum – the average rate of return from government investments (Grimsey and Lewis 2004). In 2003 the UK changed its PFI discount rate to 3.5 per cent to reflect society's time value of money (inflation). Coincidentally, Australia also changed its discount rate to a flexible one, based on the perceived level of risk on each project (the extent to which costs could escalate and erode the real value of money in terms of the physical assets it buys).

The example given here is for investment appraisal but the same approach could be used for the development appraisal of a new building or an infrastructure project. A large joinery company is considering the purchase of machinery for the production of prefabricated roof trusses for low-rise housing. The capital cost of the equipment and installation is $0.5 million. The annual income from the venture depends on the utilisation of the machinery which is in turn dependent on the state of the house-building industry. The house-building industry is currently on a major upturn. Confidence is returning after a period of recession but there are political and economic uncertainties and there is, as yet, no clear evidence that the upturn will be sustained. On this basis the managing director has calculated three scenarios as indicated in Table 3.6 where the annual incomes are net of all costs of production.

The firm normally uses an 8 per cent discount rate in investment appraisal which represents its view of the long-term cost of capital (inflation adjusted). The project NPVs were then calculated on this basis for the three scenarios and an assessment of the probability of each of the three outcomes was made. This information is displayed in Table 3.7.

On the basis of this information, the firm can calculate its ENPV as follows:

Steady	$142,680×0.6	$85,608
Growth	$280,200×0.2	$56,040
Decline	$14,420×0.2	$2,884
ENPV		$144,532

It should go without saying that ENPV carries the same advantages and limitations as all EMV methods. These are discussed in Section 3.2.4.3.

3.2.4.6   Risk-adjusted discount rate (RADR)

This is an intuitive and very simple method of dealing with risk, which is commonly used in banking and business for investments that produce an income stream over a period of time. The method is not well understood in construction but could be a very useful way of dealing with both risk exposure and risk attitude, especially for life cycle costing decisions and revenue/cost flows in PFI and PPP projects. As we said above, it is in fact a method used by the Australian Victorian Government in assessing PPP project proposals. The RADR works by gradually changing the discount rate to take account of the normal risk encountered in a development. Each increase in the discount rate effectively sets a higher hurdle for the project, making it less desirable by reducing the calculated NPV of future income. For example, a decision-maker may classify projects into low, medium and high risk with respective adjustments to the real discount rate of 1 per cent, 3 per cent and 5 per cent, respectively. The real discount rate is called the risk-free discount rate and reflects what is expected to happen in reality. A more risk averse person might set the bar higher by deciding that low, medium and high-risk projects should receive a discount rate adjustment of 3 per cent, 5 per cent and 7 per cent, respectively. Conversely, a risk seeking person might set the bar lower using discount rate adjustments of 0 per cent, 1 per cent and 3 per cent, respectively.

Of course, the opposite adjustment applies to possible *cost* streams because an increase in discount rate effectively reduces the net present cost (NPC) of future cost flows, lowering the bar and reflecting greater risk-taking attitudes. In summary, the RADR technique effectively involves making changes to the future cost and revenue flows from an investment to reflect one's risk attitude towards it. By reducing the present value of future cash flows and increasing the present value of future cost flows (safe figures), one reflects a preference for risk avoidance and by increasing the present value of future cash flows and reducing the present value of future cost flows, one reflects risk-taking behaviour.

There is some debate whether the same discount rate should be used for cost and revenue streams. For example, Grimsey and Lewis (2004) point out that costs are normally less risky than revenues, particularly in PPP projects where revenue streams are dependent on the quality of service offered. If this is the case, then revenue streams should be assessed at a higher discount rate than cost streams.

To illustrate the practical value of RADR, consider the following example of a decision of whether to invest in a new four-year lease. Let us take the steady state risk-free discount rate as 7 per cent. The normal bank premium for lending to you is, say, 2 per cent. Since the new lease is seen as a natural extension of the firm's current premises it attracts a low risk adjustment of a further 1 per cent. Table 3.8 illustrates the results of the RADR appraisal and indicates that given the above risk loading, the project does not produce a positive NPV and, therefore, should be abandoned or renegotiated.

The advantage of RADR is that it offers a way of simultaneously taking into account both risk attitude and risk exposure. The technique is also easy to understand and calculate. However, its disadvantages flow from the lack of any explicit methods for calculating the risk adjustments which tend to be rather informal and difficult to justify on any logical basis. Furthermore, risk adjustments in RADR take account of both risk exposure and risk attitude, it is not feasible to disentangle one from the other.

3.2.4.7   Simulation

We mentioned earlier that the problem with single point estimates, such as those used in the majority of construction projects, is the potential variability that they hide. In reality we know that estimates can vary depending on a whole range of events, each of which has its own probability of occurring – some escalating the cost, some reducing the cost and some not affecting it. We use the term "probability distribution" to describe the sum of an estimator's knowledge about the full range of upside and downside risks that can affect a decision or a project.

Simulation is a sampling technique that takes this into account *on* a holistic project level. It randomly draws values from the full range of individual probability distributions developed for each decision on a project, providing the systematic evaluation of alternative project strategies and outcomes and the search for the optimum one. Traditionally, the *Monte-Carlo* technique is used as the statistical basis for such analysis. Although many managers

Table 3.8 RADR appraisal

RADR$_1$ (benefits) = 7 + 2 + 1 = 10%
RADR$_2$ (costs) = 7 − 2 − 1 = 4%

Year	Gross income ($)	RADR = 10%		RADR = 4%	
		DCF ($)	Costs ($)	DCF ($)	NCF ($)
0			(500,000)	(500,000)	(500,000)
1	580,000	527,220	(400,000)	(384,800)	142,420
2	600,000	495,600	(400,000)	(370,000)	125,600
3	600,000	450,600	(400,000)	(355,600)	95,000
4	600,000	409,800	(400,000)	(342,000)	67,800
NPV (risk adjusted)					(69,180)

have heard of this simulation technique, it conjures up images of a complex analytical tool that is difficult to use. Nevertheless, the Monte-Carlo technique is quite simple in principle and recognises individual variables within a calculation as probability distributions rather than single numbers. By using Monte-Carlo simulation, probability distributions for any decision (as defined by the estimator) can be randomly combined using random numbers (much in the same way as a roulette wheel – thus the name Monte-Carlo) to produce a complete judgement about the entire range of potential events. This produces a multi point estimate reflecting the likelihood of each value in that range. Using a simulation program (probably built on the back of a spreadsheet such as Excel) a project is "built" many times, with random variations of the input variables defined in the input probability distributions for each decision in a project. The simulation results in a statistical sample of different project outcomes with identical probabilistic characteristics. Analysis of this sample enables us to attach some numeric evaluation to the degree of risk in an estimate.

For example, a manager has to decide on a bid for a project and Figure 3.4 represents the estimator's range of possible outcomes (probability distributions) for each element of the estimate.

The estimate was simulated 1000 times using Monte-Carlo analysis and the following combined outputs produced (Figures 3.5 and 3.6) which

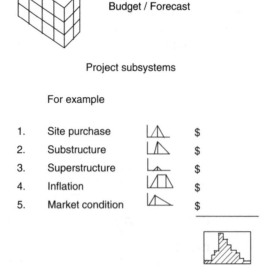

Figure 3.4 Probabilistic forecast.

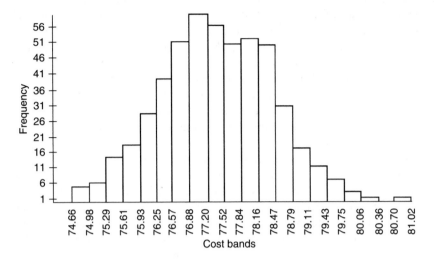

Figure 3.5 Probability distribution.

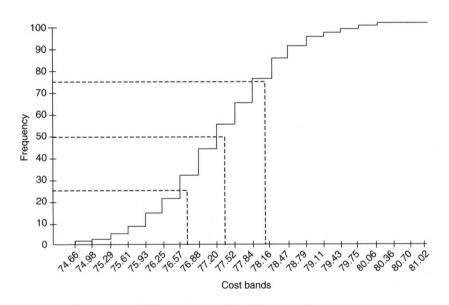

Figure 3.6 Cumulative probability distribution.

makes use of the estimator's full range of knowledge concerning the most likely, best and worst cases for each component of the project.

The cumulative probability distribution in Figure 3.6 is an extremely powerful yet very simple tool for estimating the risk associated with any bid price. For example, it is evident that there is a 100 per cent chance that the project cost will not exceed $81.02 million which means that a bid of this level attracts no risk (that the project will exceed this amount if it was won). It does not of course reflect the risk of losing the tender against competitors in a bid (although a probability distribution of this could also be constructed to indicate this). In contrast, there is a 100 per cent chance that the final cost will exceed $74.66 million, which means that a bid of this amount will attract a 100 per cent risk (that the project will exceed this amount if it was won). Knowing this and the relative probabilities of costs exceeding budgets by different amounts, it is possible for a decision-maker to know exactly what level of risk is being taken on any bid price. In contrast, when estimates are produced deterministically (without probability distributions) rather than stochastically (with probability distributions) as is often the case in practice, it is not possible to make this judgement about risk. Indeed, it is interesting that when deterministically derived bids are put through simulations, it is normally the case that they turn out to be very highly risk averse (overestimated costs and underestimated revenues), meaning that bids have been needlessly lost or that prices for customers are needlessly high.

Nevertheless, traditionally, in construction and property development, estimates and forecasts of, for example, time, costs, prices, rents and returns have been given as single deterministic figures. While there are clearly limitations with this approach, there are sometimes very good reasons for this. For example, single point figures are often needed as budgets for the purposes of planning and control. Most projects need to be managed so that they are completed by a specific date and on or under a specific budget. In such cases, single point figures are necessary as targets for the purposes of day-to-day planning and control. Nevertheless, it remains the case that these single points are forecasts and targets, not certainties. For most purposes, single point forecasts conceal more than they reveal.

The advantage of simulation is that the resultant probability density function and cumulative distribution functions are a powerful yet simple measure of project risk exposure. The results also allow for the informal incorporation of risk attitude by the decision-maker who has to decide the level of risk to accept as part of a valid forecast. Simulation also requires very little knowledge of mathematics and can cope with a large number of types of input distribution. This gives flexibility to model precisely, the perceptions of uncertainty surrounding all of the input variables. It is also possible to deal with correlation between components of the analysis although this requires more sophisticated software (see Section 3.10). The disadvantage of simulation is that it relies on the use of a computer. Software has to be purchased or developed and analysis needs to be undertaken and structured skilfully in order to decompose the problem into relatively independent sub systems or to take into account the correlation between them. Simulation also requires the construction of probability distributions for each decision variable, which requires some knowledge of how to construct probability distributions, using techniques such as the Delphi method. Finally, it is also necessary to choose an appropriate distribution type to input into the simulation.

CHOICE OF INPUT DISTRIBUTIONS FOR SIMULATION

In practice, almost every decision is subject to a different range of upside and downside risks. This can be represented in a triangular probability distribution (Figure 3.7 – the most likely value is "b", the lowest possible cost is "a" and the highest possible cost is "c"), assuming that the variable is "cost". The distribution is skewed to the right, which indicates that downside risks (events that produce costs increases) are greater than upside risks (events that can reduce costs). Left hand skews such as Figure 3.8 are produced when the opposite is true. In this case, there is no chance of the cost exceeding b. On the other hand, some decisions have an equal probability of a whole range of costs which needs to be reflected in a uniform distribution (Figure 3.9). In contrast, Figure 3.10 represents a combination of both uniform and triangular distributions (trapezoidal distribution), Figure 3.11 is a combination of uniform distributions (step-distribution) and Figure 3.12 is a distribution with only two possible values (discrete distribution).

Normally the choice of distribution is not based on a search for the "true" distribution for a variable but on the objective of modelling the estimators' perception of the range of outcomes and associated probabilities. It is always important to keep one's "feet on the ground" and remember that without accurate statistical data, which is almost always the case in industries like construction, we are in the realm of subjective decisions of outcomes and probabilities. This is why the distributions we have identified for use above

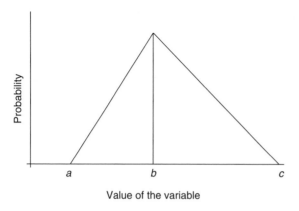

Figure 3.7 Triangular distribution – right skew.

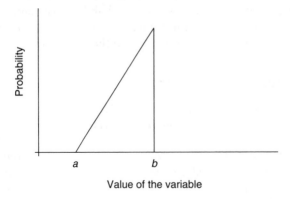

a   b

Value of the variable

Figure 3.8 Triangular distribution – left skew.

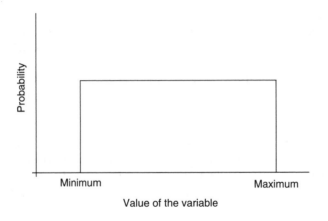

Minimum   Maximum

Value of the variable

Figure 3.9 Uniform distribution.

are simple. In reality, most estimators will be able to state reasonably clearly that the cost or time for a particular variable will not exceed x or be less than y and what the most likely outcome will be. Furthermore, the apparent simplicity of these distributions belies their power since they have been shown on many projects – including undersea oil and gas and large defence projects – to produce robust models of risk.

CORRELATION IN SIMULATION

When considering risks in project programmes, when one event could depend on the outcome of another, correlation between the different probability

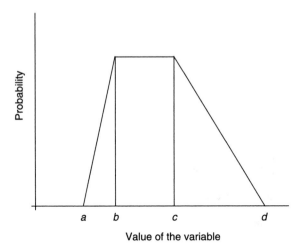

Figure 3.10 Trapezoidal distribution.

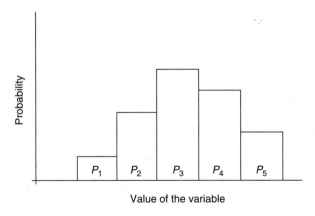

Figure 3.11 Step distribution.

distributions for each event can become important. For example, assume that activity P is dependent on the outcome of an earlier activity K. For activity P, a number of different distributions are entered, each one contingent upon a specific result from activity K. Now, in the simulation, when it is time to draw a number for activity P, the programme checks back to read the result drawn for activity K during the same pass. The correlation is said to be positive when a good result in K implies a good result in P, or negative when a good result in K produces a poor result in P.

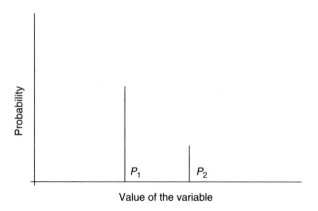

Figure 3.12 Discrete distribution.

Some of the larger and more expensive simulation software packages can deal with correlation – the detail of how strongly the links are defined being left to the user. However, it is the case that outside the aerospace, defence, oil and gas industries, the majority of projects can be adequately simulated using very simple and inexpensive software. For instance, compared to aerospace projects, construction projects are far less heterogeneous and risky and do not justify such complexity.

3.2.4.8 Stochastic dominance

Comparing projects on the basis of single point deterministic estimates is a straightforward task. If, however, the project has been estimated using a simulation approach to produce a probabilistic result, then the task of comparing one project with another becomes more complex. Sometimes by superimposing the probability density functions (PDFs) and the cumulative distribution function, it will be clear which project has stochastic dominance (superiority in terms of probability distributions). For example in Figure 3.13 it is clear that Project 1 has stochastic dominance over Project 2 since it has a lower expected (mean) cost. It is also apparent from Figure 3.14 that for any chosen level of life cycle cost, there is a higher likelihood that it will be achieved on Project 1 than on Project 2.

The comparison of Projects 1 and 2 is straightforward since the two projects have an identical variance or distribution about the mean (sometimes called the standard deviation). This reflects the chance (risk) that the cost will vary, either positively or negatively, from the expected (mean) cost. However, in Figure 3.15, Project 3 and Project 4 have different variances which makes comparisons more difficult. Here, Project 3 has both a lower mean cost and

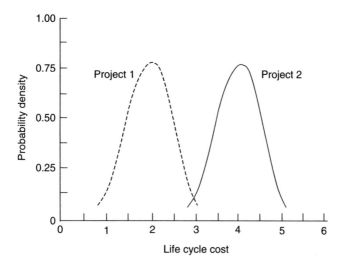

Figure 3.13 Probability density function of life cycle costs for Projects 1 and 2.

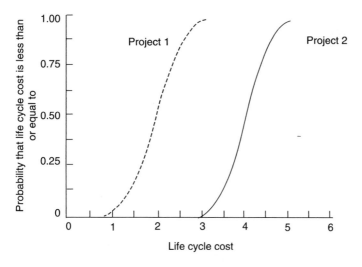

Figure 3.14 Cumulative density function of life cycle costs for Projects 1 and 2.

a lower variance which means it is better than Project 4 in both ways. This is reflected in Figure 3.16 in the relatively steep slope of the cumulative distribution curve of Project 3, which illustrates that for the majority of life cycle costs (over $1.5 million), the probability of achieving it is higher in Project 3 than Project 4.

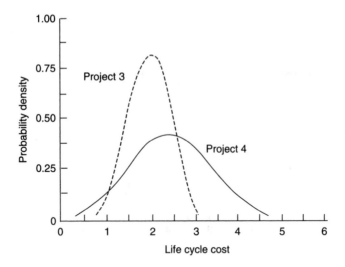

Figure 3.15 Superimposed probability density functions of life cycle costs for Projects 3 and 4.

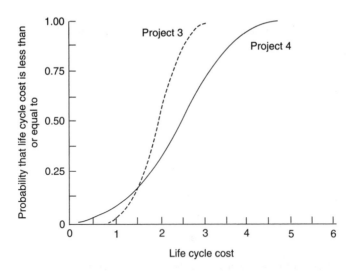

Figure 3.16 Superimposed cumulative density functions of life cycle costs for Projects 3 and 4.

Stochastic dominance is not really an analysis technique. It is more – a method of assessing the output of analysis – and its principal advantage is that it offers a way of comparing probabilistic results. The method is largely informal and the results may depend on personal risk attitude, which is exercised informally by viewing the graphs.

3.2.5 Risk management software

The rapidly changing business environment, the vast amount of data which is generated on projects and increasingly stringent corporate reporting requirements have heightened the need for better management systems that can record, store, organise and facilitate rapid access to risk-related information. In addition, the development of national standards for risk management such as the Australian and New Zealand Standard for Risk Management, AS/NZS 4360: 1999, have provided a widely accepted model for risk management practice and, in doing so, have created a market and catalyst for the development of many new software packages.

However, a sense of perspective must be maintained. It is certainly not the case that the latest risk management software is an essential tool for every risk management programme. To think that risk management can be successfully practised with the press of a few buttons on a keyboard is not only dangerous but grossly oversimplifies the risk management process. It is important to remember that risk management is actually more a philosophy than simply a process of buying some off-the-shelf software. Furthermore, software plays little or no role in the formulation of control strategies to respond to risks. The only ingredient that is absolutely necessary for good risk management practice is an appropriate attitude of mind. Ultimately, effective risk management is dependent on well-trained, highly experienced human experts who are able and encouraged to think creatively and imaginatively about a range of future project outcomes attached to a wide variety of events related to a project and its political and economic environment. Organisations which do not have the resources to invest in expensive software and computer systems can be just as effective, if not more effective than wealthy organisations, in managing their risks and opportunities. Nevertheless, having said this, when used appropriately, software can be a useful tool to assist risk analysis and it is important to know how to approach the rather confusing and intimidating software market.

In approaching the software market, the important criterion is to consider where the maximum benefit lies. This will not necessarily be correlated to the price paid for the software product. Good work in risk analysis can be done with very basic and inexpensive software, accompanied by thorough training of relevant staff. In general, software costs will start to escalate if there is a requirement to ensure that the risk analysis tool produces standardised reports in a corporate style for the firm in question and which communicate with information in the firm's other management information systems. Integrated risk management systems, which have this interoperability capability, are at the cutting edge of current developments in risk management software.

Essentially, there are two types of risk-related software products on the market: *risk management information system* (RMIS) (software which is designed to manage the entire risk management process) and *risk analysis software* (software which is designed to facilitate only the risks analysis stage of the risk management process). These products may be purchased separately or combined and, while interdependent, have quite different functions which are reviewed below.

3.2.5.1   Risk management information system software

In contrast to risk analysis software, which specialises in supporting the risk analysis process, RMISs are designed to provide a more general means to support the entire risk management process. There is, of course, an inter-relationship because an RMIS should provide a mechanism for collecting and disseminating data for reliable risk analysis. This requires a facility to enable uniform record keeping in key areas of risk such as occupational health and safety incident recording, insurance and workers compensation, environmental impacts, quality management, claims management etc. In addition to supporting risk analysis, RMISs are meant to provide easy access to critical performance data. They should also facilitate reporting to man-agement and stakeholders on a range of KPIs generated internally or by statutory reporting, auditing and licensing requirements. RMISs are there-fore useful in enabling organisations to more easily demonstrate compliance with state and national legislation, standards and codes of practice. Furthermore, they can be useful in providing tangible evidence to directors and senior managers that the organisation is meeting its obligations to its employees, shareholders, the public and the environment.

When choosing RMIS software, it is important to realise that there are essentially two types of product, namely *process driven* and *information driven*. Process driven software is designed to help decision-makers work their way through a logical process that is built into the program. The quality of this type of program is dependent on the quality of the processes hard-wired into the software. In contrast, information driven software imposes no particular discipline on users but merely provides a database of risk-related information which can be accessed by users. The quality of this type of system is dependent entirely on the database design and reporting tools provided.

3.2.5.2   Risk analysis software

Risk analysis software is designed to overcome the complexity and tedium that is often involved in quantitative risk analysis. However, before pro-gressing to a discussion of how one approaches the risk analysis software market, it is important to remember that risk analysis plays a very small part in the overall risk management process. Indeed, there will frequently be situations where formal risk management is practised and a quantitative analysis is never undertaken. A thorough risk identification stage in itself may well lead to the generation of sufficient information for the decision-makers to take action without a formal quantified analysis. Alternatively, decision-makers may opt for a qualitative analysis and never progress to a quantitative analysis. So risk analysis software may not be necessary and even when it is, the number crunching part of a simulation may well be the least important part of the total time spent building the assumptions and distributions which form the basis of it. Nevertheless, when a quantitative risks analysis is appropriate, there is little doubt that good risk analysis software can be a great help in expediting the number crunching part of the risk analysis process.

The market for risk analysis software is small and specialised. Risk analysis is a field that requires significant training and support and many software houses find it uneconomic to develop and sell packages. Furthermore, some software has been introduced before it has been well tested resulting in frustrating problems for many users. Indeed, some products have been withdrawn altogether, not only due to poor design but also due to a lack of interest among buyers.

Essentially, there are four main types of risk analysis software on the market. The first is the custom-designed product. This ensures that the software fits into the in-house project management system. The second is to consider the purchase of relatively cheap add-ons to spreadsheet programs. These usually cost per user and rely on the computer user already having a spreadsheet program running. An example of this type of program is @Risk, which links to Excel and is probably the most widely used software in the construction industry. The third approach is to look for relatively large, off-the-shelf packages which are usually based around project management suites such as ARTEMIS and PRIMAVERA. These are extremely powerful project management tools, add ons enabling users to make probabilistic inputs to network activities to deal with both time and costs. Correlation analysis is also possible. The disadvantage of these programs is their complexity, partly because they are so powerful. They are difficult to use and also require a project to be set up as a critical path, which is not always appropriate, especially for smaller projects. The fourth approach is to consider a software product – which may be stand-alone or may double as a project management tool similar to those mentioned. These are often developed by consultants who specialise in risk, rather than a software house. These products tend to be excellent but are very expensive.

3.2.5.3 Software risks

In the past, when most software was custom-made and manufactured from scratch, organisations had full control over the requirements, design and integration of their new IT systems and softwares. In contrast, today's managers are faced with choosing from a limited range of off-the-shelf

IT and software components from suppliers who are under pressure to continually upgrade and enhance their product capabilities and interfaces. While this has some advantages, the lack of control over the design process means that managers must try and get the best fit between their requirements and the systems available. Unfortunately, because each product has its own idiosyncrasies, this often means significant integration problems revolving around incompatible operating systems and unstable interfaces. Given the increasing dependence of organisations on IT, problems like this, of software integration, are becoming one of the greatest risks facing organisations and the choice of appropriate software becomes a critical consideration.

To minimise these risks, it is critical to have a strategy for selecting the best software for your business. In assessing this, it is important to consider a number of issues, which include the types of projects your are involved in, the risk management maturity of your organisation, the skills and attitudes of your workers towards risk management, the amount of money and time you have to spend, what your exact requirements are and how prepared and capable the organisation is to maintain it. Generally, speaking, an organisation involved in larger and more complex projects will justify more complex software. For example, the most complex softwares would be suitable for companies involved in off-shore oil and gas installations and petrochemical plants etc. Those firms involved in the PPP and PFI market may also consider more sophisticated software which is able to model risks longitudinally over periods of time. The risk management maturity of an organisation is an important consideration because there is little point investing in software if basic of risk management is not being practised, is not understood and is not supported by top management. There are many examples of failed risk management systems that have been developed and implemented by well-meaning and knowledgeable professionals without the support of top management. Immature organisations also will not have the data available to input into the analysis process, so investment in a quantitative analytical tool would be pointless. A positive attitude towards risk management and some analytical skill base are a basic prerequisite before putting aside resources for investing in risk analysis software. For example, conditional branching is one of the indicators of top-quality software in this field. Nevertheless, having access to the functionality of conditional branching requires that the user has quite a sophisticated understanding of the issues surrounding correlation between risk variables and probability distributions. Handing a system with this level of functionality to someone who does not understand these issues would be like placing a high powered motorcar in the hands of a novice driver. The driver is actually likely to get from A to B more safely and quickly in a less powerful car. Finally, an important issue which is often neglected when selecting appropriate software is to clarify exactly how prepared and capable an organisation is to maintain it. The best decisions are those where the capabilities of the software match the needs and capabilities of the organisation. No matter how sophisticated the software, if the business environment does not support it, then it will not be implanted effectively. By not considering this, many organisations have made the mistake of purchasing overly complex software which needs to be fed continuously with data, meaning that it becomes a liability rather than an asset.

3.3 Quantitative versus qualitative risk analysis

Because modern society places great faith in numbers, there is considerable pressure on managers to look to statistical calculations to justify their decisions. After all, complex calculations based upon statistics collected elsewhere can provide a convincing, convenient and almost unchallengeable façade behind which managers can hide. For these reasons, many managers believe that if one is not doing quantitative risk analysis then one is not doing risk analysis at all. This perception is wrong and potentially dangerous. First, reliable quantitative risk analysis is not always possible because of difficulties in attaching reliable statistics to many types of risks (particularly those which arise from human sources). Indeed, it is rare to find a situation where the numerical data available to decision-makers is any more than an estimate. It may be a good estimate, but an estimate nonetheless. In this situation, playing mathematical games with intrinsically inaccurate data will not only incur unnecessary costs and time but may compound basic inaccuracies, encourage undue confidence and increase the risk exposure of the organisation. While numbers might be very valuable in analysing some types of risk, say in financial and engineering problems where variables are quantifiable, they can be misleading with others. Quantitative risk analysis is only as reliable as the data upon which it is based and there is a real danger that the allure of a calculable outcome, especially when embodied in a comprehensive computer package can focus attention on those factors handled within the package to the detriment of larger factors outside it. Trying to force quantitative methods on a qualitative problem with the sole purpose of imposing some veneer of mathematical respectability is futile. Indeed, it is not even necessary. A common law requirement to assess risk so far as it is *reasonably practicable* does not expect or require managers to rely upon numbers to this extent. For example, compliance with the UK's CDM regulations, Australia's Occupational Health and Safety Regulation 2001 and its ASX Corporate Governance Guidelines can, in most instances, be satisfied by qualitative analysis.

Broadly speaking, you should conduct a quantitative risk analysis only:

- When you have first conducted a qualitative analysis
- On the risks which emerge as particularly important from qualitative analysis

- When you have sufficient time or when time becomes available
- When you have reliable data to assign numbers to probabilities or when data becomes available
- When it makes sense to attribute numbers to the consequences of a risk
- When you have the necessary expertise or support to conduct and interpret the results of a quantitative analysis.

Examples of situations where a quantitative analysis may be appropriate and useful include:

- Estimating the cost and contingency allowances for a contract
- Plant reliability assessment
- Life cycle costing
- Economic appraisal of a new business opportunity.

The key point is that risk analysis needs good information but this does not necessarily mean numerical information. This has been supported by the many highly successful organisations which go no further than qualitative risk analysis. To them, the primary benefit of risk management is in forcing people to set job priorities and to think through projects in more detail.

3.4   Qualitative risk analysis

The basic principle that underpins both qualitative and quantitative risk analyses is exactly the same. It is that every event that represents a risk to an organisation has a *probability* of occurring and a *consequence* if it occurs. The magnitude of the risk is its:

Probability × Consequence
(*likelihood*) (*impact*)

The only difference between quantitative and qualitative analysis is the way that values are attributed to these two components. In quantitative analysis, numeric values are used but in qualitative analysis, words (or descriptors) are used. An example of descriptors commonly used to qualitatively label probabilities, consequences is shown in Tables 3.9 and 3.10.

The labels used in Table 3.10 are used on many, if not most, companies's risk management systems. However, there is a major problem in that these labels do not adequately recognise the upside (opportunistic) aspect of risk. For example, the idea of a catastrophic opportunity makes no sense! Therefore, despite the rhetoric of many companies's systems documentation, which allude to recognise the upside of risk, the vocabulary in most systems does not facilitate this. This is not surprising since most texts on risk management focus on the downside of risk and have never recognised this other

Table 3.9 Qualitative probability labels

Descriptor	Description
Rare	This event may occur in exceptional circumstances only
Unlikely	This event is not likely to occur
Possible	This event could occur at some time
Likely	This event has happened before and will probably occur again
Almost certain	This event is common and is expected to occur in most circumstances

Table 3.10 Qualitative consequence labels

Descriptor	Definition
Insignificant	No injuries, low financial loss
Minor	First aid treatment, on-site release immediately contained, medium financial loss
Moderate	Medical treatment required, on-site release contained with outside assistance, high financial loss
Major	Extensive injuries, loss of production capability, off-site release with no detrimental effects, major financial loss
Catastrophic	Death, toxic release off-site with detrimental effect, huge financial loss

Source: AS/NZS 4360: 1999.

dimension. Here are some examples, spanning a range of industries, which have undoubtedly influenced the development of many company's systems – PMI (1992), Boothroyd and Emmett (1996), ICE (1998), Bowden *et al.* (2001) and FMA (2004). Each of these influential texts distinguishes between risks and opportunities but then goes on to recommend a vocabulary and set of control mechanisms designed to *mitigate* rather than *optimise*. It is ironic that most customers come to the private sector for innovation – yet many company systems are set up to discourage it. To those who work in these companies, risk management can be a defensive and potentially recriminatory process whereas it should be an exciting, positive, entrepreneurial and value-adding process.

We have argued throughout this book that risk management, if seen positively, can make an enormous contribution to optimising an organisation's triple bottom line – its social, environmental and financial performance. To foster this opportunistic culture, we recommend using the consequence labels in Table 3.11 rather than those in Table 3.10. Table 3.11 distinguishes between risks and opportunities, provides more detailed descriptions for different levels of significance and replaces the word catastrophic with extraordinary. Separate descriptors for risk and opportunity are important because there is considerable evidence to indicate that decision-makers have great difficulty in thinking about risks and opportunities simultaneously and that they need to be managed separately (Loosemore and Lam 2004).

Table 3.11 Qualitative labels reflecting risks and opportunities

Label	Description for risks	Description for opportunities
Insignificant	Insignificant negative impact on key performance indicators. No injuries, low or zero financial loss.	Insignificant benefits to KPIs. No potential benefits to staff, employees or customers.
Minor	Minor negative impact on key performance indicators. First aid treatment required, on-site release immediately contained, minor financial loss.	Minor benefits to KPIs. Some measurable improvement possible in financial, response times, safety and environmental performance and thereby quality of service to customers.
Moderate	Significant negative impact on key performance indicators. Medical treatment required, loss of production capability, off-site releases with no detrimental effects, medium financial loss.	Significant benefits to KPIs. Significant possibility of improvement in financial performance, quality of service or in the functional, physical and financial performance of a contract.
Major	Major negative impact on key performance indicators. Extensive and severe risk of economic and financial loss, legal or industrial action, to the health and safety of staff or the general public, to the environment or to corporate image.	Major benefits to KPIs. Could add substantial extra value to customers' business performance, exceeding customer expectations and leading to a major advance in corporate image, safety, financial, quality and environmental performance.
Extraordinary	Catastrophic impact on key performance indicators. Major incident involving fatalities, environmental disaster, bankruptcy, shareholder losses, public image.	Enormous opportunity to increase quality of service to customers beyond all expectations, regulations and best practice. Setting new global benchmarks across all KPIs, standards of service and creating new business opportunities, enhanced public image and community perceptions.

Having developed qualitative matrices for probabilities and consequences, it is now possible to estimate the level of risk (probability × consequences) by producing a combined matrix as illustrated in Table 3.12. Using this is very easy. For example, if an event has a "moderate" probability of occurring and a "major" consequence if it did occur, the risk level according to Table 3.12 would be "high".

Table 3.12 Qualitative risk estimation

Probabilities	Consequences				
	Insignificant	*Minor*	*Moderate*	*Major*	*Extraordinary*
Almost certain	Low	Medium	High	High	High
Likely	Low	Medium	Medium	High	High
Possible	Low	Low	Medium	High	High
Unlikely	Low	Low	Low	Medium	Medium
Rare	Low	Low	Low	Medium	Medium

Key

High – are those risks or opportunities with a relatively high likelihood and large impact. They will require close management attention at senior level, detailed research, quantitative analysis if possible and a formal risk action plan. Any management strategy should be developed in close consultation with senior managers and a Risk Manager.

Medium – are risks or opportunities with a medium likelihood and impact. You should decide how they are to be managed in consultation with a Risk Manager who will decide whether senior management attention and further quantitative analysis is necessary. A formal risk action plan is normally needed which clearly allocates responsibilities and timetables for action.

Low – are those risks or opportunities with a relatively low likelihood and impact. They are regarded as acceptable within normal business activities, are managed effectively by routine and standard procedures and can be excluded from further detailed consideration. However, they cannot be ignored because if standard procedures and controls do not work, a minor risk can quickly become a major problem or a potential opportunity may be missed.

There are a number of important and potentially costly mistakes which companies make when developing and using their risk matrix. First, many companies fail to define precisely what they mean by a high, moderate or a low risk. For example, a company may decide that a "high" risk demands senior management attention and detailed quantitative analysis but that a "moderate" risk does not. If this is not made explicit then it is likely that a decision-maker will fail to understand the significance of the event they have identified and to respond appropriately. A second common mistake is not to think carefully about the location and relative numbers of high, medium and low categories in the risk matrix. Many companies fail to appreciate that the positioning of the high, medium and low categories within their risk matrix is an important consideration which is a reflection of their risk attitude and appetite. For example, a company that is risk averse should have more "high" categories within their matrix than one which is risk seeking. If this is not discussed in depth, then the company could inadvertently lose business opportunities which are well within their capabilities and desired risk portfolios.

Despite the above problems, the advantage of qualitative risk assessment is that most competent managers should find it quick and easy to master. It is also something that can be realistically achieved in most organisations with minimal training. Furthermore, it is appropriate where risks cannot be meaningfully quantified and can provide managers with a general understanding of comparative risks associated with different events which can be prioritised into distinct risk categories. This of course is assuming that the process is well structured and managed. However, qualitative risk analysis also has a number of disadvantages. Apart from the potential problems discussed in the previous paragraph, it is quite imprecise because the categories of probability, consequence and risk are generally quite crude. Furthermore, different risk events placed within these broad categories can have very different levels of risk. Inevitably, categorising an event in terms of its likelihood and consequences is a highly subjective process which is difficult to justify. This means that there is no guarantee that the different quadrants of "high" risk in Table 3.12 actually represent the same level of risk. The consequence is that comparisons between different risk classes can often lead to inconsistencies. Another problem with qualitative risk analysis is that it is difficult to compare different risk events on a similar basis such as a dollar value. Furthermore it is difficult to associate a qualitative risk assessment with an appropriate and economically viable choice of risk treatment options. For example, the output of qualitative risk assessments cannot be incorporated into techniques such as cost-benefit analysis, which can be very useful in identifying economically viable risk response options.

3.5   Semi-quantitative risk analysis

To overcome some of the disadvantages of qualitative risk analysis, it is possible to employ a semi-quantitative analysis. This takes qualitative risk analysis a step further by attributing predefined values to the probability and consequences labels which can then result in more refined and precise estimates of risk which can be used to adjust programmes, estimates or bids. For example, in Table 3.13, a company has sat down and discussed exactly what the different labels mean to them and quantified them accordingly. This example relates

Table 3.13 Semi-quantitative analysis using absolute values

Likelihood	Consequence values ($)				
	Insignificant <$10,000	Minor $10,000–$30,000	Moderate $30,000–$60,000	Major $60,000–$90,000	Extra ordinary >$90,000
0.90 (Almost certain)	<9,000	9,000–27,000	27,000–54,000	54,000–81,000	>81,000
0.70 (Likely)	<7,000	7,000–21,000	21,000–42,000	42,000–63,000	>63,000
0.50 (Moderate)	<5,000	5,000–15,000	15,000–30,000	30,000–45,000	>45,000
0.30 (Unlikely)	<3,000	3,000–9,000	9,000–18,000	18,000–27,000	>27,000
0.10 (Rare)	<1,000	1,000–3,000	3,000–6,000	6,000–9,000	>9,000

to the risk of a cost overrun on a $500,000 project and allows a company to compare the risks within projects and across different projects on a commonly agreed scale. Without such a scale, different people would attribute different values to the various probability and impact levels according to their own risk attitude – not necessarily the company's. Furthermore, the figures in the table cells give some rough indication of what single point contingency allowances would need to be made for each level of risk.

Of course, every company will attribute different numbers to the probability and consequence labels, resulting in different values within the table. For example, a company which is highly risk averse would be likely to attribute relatively higher numbers to the probability labels and smaller numbers to the consequence labels. A smaller company with fewer resources would tend to do the same. For such a company, a $50,000 loss might be extraordinary and bring it to the verge of bankruptcy while for a large firm this may be a relatively insignificant loss. Clearly, companies need to adapt the risk matrix to suit their won financial horizons. Some companies might even use different matrices for different projects, if they vary enough in value.

To overcome the inconvenience of producing numerous risk matrices for different projects or types of risk, some organisations use weightings rather than precise numbers. In this approach to semi-quantitative analysis, the values attributed to different categories of likelihood and consequences reflect the relative magnitude of consequences and likelihood rather than absolute values. This is illustrated in Table 3.14 where company discussions have resolved that the consequences of a "major" event (say, an injury) are twice those derived from a "moderate" injury. The resultant risk factors within the table itself then provide a much more accurate scale of risks and can become weightings that are applied to estimates or forecasts to reflect relative levels of risk. It is interesting to note that the shading in Table 3.14 indicates a high risk taking

Table 3.14 Semi-quantitative analysis using relative weightings

Likelihood	Consequence weightings				
	Insignificant 0.05 (5% of project costs)	Minor 0.10 (10% of project costs)	Moderate 0.20 (20% of project costs)	Major 0.40 (40% of project costs)	Extra ordinary 0.80 (80% of project costs)
0.90 (Almost certain)	0.045	0.09	0.18	0.36	0.72
0.70 (Likely)	0.035	0.07	0.14	0.28	0.56
0.50 (Moderate)	0.025	0.05	0.10	0.20	0.40
0.30 (Unlikely)	0.015	0.03	0.06	0.12	0.24
0.10 (Rare)	0.005	0.01	0.02	0.04	0.08

Key
High risk >0.35 control and produce detailed risk management plans.
Medium risk 0.10–0.35 produce contingency plans should it eventuate.
Low risk <0.05 monitor risk and produce outline contingency plans.

organisation because of the relatively few high-risk cells. An alternative explanation is that the organisation is operating in a relatively low risk environment.

Semi-quantitative analysis overcomes some of the disadvantages of qualitative analysis. For example, it is more useful in a practical sense and provides a useful basis for understanding comparative risk levels between different events. It also results in a less crude categorisation of risks (not just high, moderate and low). In Tables 3.13 and 3.14 we have a whole range of numbers reflecting a far wider range of risk levels. Nevertheless, the numbers attributed to the various categories of risk and consequences are often meaningless (particularly in consequences). Numbers are useless unless people think about what they mean and about their relativity. At best they are still subjectively derived estimates (whether they are relative or absolute values), which means that it is not possible to say that two events that are assigned the same risk values do in fact represent the same level of risk. Furthermore, without producing different matrices for different types of decisions, the analysis becomes even cruder. Finally, it is always important to remember that if numbers cannot be sensibly attributed to probabilities and consequences, and if the resulting risk indices cannot be used to adjust estimates or programmes, then there is no point in undertaking a semi-quantitative analysis.

3.6   Conclusion

In this chapter, we have discussed the science of risk analysis. Our aim has been to dispel the widely held impression that reliable risk analysis always involves complex statistical procedures and computer programs. Using numerous practical examples, we have introduced basic concepts of probability, showed how probabilities are best derived in different contexts and illustrated some valuable quantitative and qualitative techniques. We have also tried to introduce the idea of risk attitude as a potential source of bias in the risk assessment process. In the next chapter, we continue this theme and discuss, in more detail, the many psychological processes which can introduce human bias and error into the risk management process. However, we end this chapter with two case studies which illustrate the practical application of the techniques discussed above. One looks at the risk analysis process on a large land reclamation scheme in China. The other explores the risk analysis process on a new bridge project in central London.

3.7   South China land reclamation

In this section we explore the risk analysis process on a large land reclamation scheme in Shenzen in Southern China. Shenzen is a special economic zone in the province of Guangdong and, due to buoyant economic growth, is facing shortages of land along the coast. For this reason, a land reclamation scheme is proposed to provide the space needed for a new container port and ancillary facilities.

3.7.1   Project scope

The land reclamation project entailed the reclamation of 250 hectares of land, the construction of sea walls, extension of existing culverts and the relocation of existing ferry piers. The summary page of the lowest tender is shown in Table 3.15.

After a meeting between the client's team and the lowest tenderer, bills 4 (Table 3.16) and 5 (Table 3.17) and the contingency sum (Table 3.18) have been identified and agreed as high risks areas that will need constant monitoring. A sensitivity analysis of each risk area, which graphically illustrates the variations identified in Tables 3.16, 3.17 and 3.18 is illustrated in Figure 3.17.

Table 3.15 Summary page of lowest tender

Description	Million (Rmb)
The lowest tender Bills of Quantities Summary	
Bill No. 1 – Preliminaries	56.00
Bill No. 2 – Site clearance	0.30
Bill No. 3 – Roadworks	22.00
Bill No. 4 – Reclamation and sea walls	343.00
Bill No. 5 – Piers	152.00
Bill No. 6 – Pumphouse and pumping mains	64.00
Bill No. 7 – Drainage culverts	37.00
Bill No. 8 – Borrow areas	52.00
Bill No. 9 – Provisional sums and dayworks	30.00
Subtotal of bills 1 to 9 inclusive	756.30
Allowance for contract price fluctuation increase	50.00
Contingency sum	73.00
Total	879.30

Table 3.16 Bill No. 4

Variation (%)	Bill sub-total (million Rmb)	Summary total (million Rmb)
+50	514.50	1050.80
+40	480.20	1016.50
+30	445.90	982.20
+20	411.60	947.90
+10	377.30	913.60
0	343.00	879.30
−10	308.70	845.00
−20	274.40	810.70
−30	240.10	776.40
−40	205.80	742.10
−50	171.50	707.80

Table 3.17 Bill No. 5

Variation (%)	Bill sub-total (million Rmb)	Summary total (million Rmb)
+50	228.00	955.30
+40	212.80	940.10
+30	197.60	924.90
+20	182.40	909.70
+10	167.20	894.50
0	152.00	879.30
−10	136.80	864.10
−20	121.60	848.90
−30	106.40	833.70
−40	91.20	818.50
−50	76.00	803.30

Table 3.18 Contingency sum

Variation (%)	Bill sub-total (million Rmb)	Summary total (million Rmb)
+50	109.50	915.80
+40	102.20	908.50
+30	94.90	901.20
+20	87.60	893.90
+10	80.30	886.60
0	73.00	879.30
−10	65.70	872.00
−20	58.40	864.70
−30	51.10	857.40
−40	43.80	850.10
−50	36.50	842.80

3.7.2 Analysis

We are studying this project at the point where tenders have been received. By carrying out sensitivity tests on the most attractive bid, we hope to minimise exposure to risk by identifying areas where our maximum project management efforts and limited resources will be directed. Each line in the spider diagram (Figure 3.17) indicates the impact on the total cost, of a defined variation in a single parameter that has been identified as having a potential impact on costs. The flatter the line, the more sensitive the total costs is to a certain variation is that parameter. Figure 3.17 illustrates that the project cost is more sensitive to the "reclamation and sea walls" element of the bid than it is to the "piers" and "contingency" elements of the bid. In others words, the same percentage of change in each of these elements produces a relatively larger effect on the total contract sum in the case of reclamation and sea walls than in any other bid element. It is in these areas that project management efforts and limited resources would be best directed.

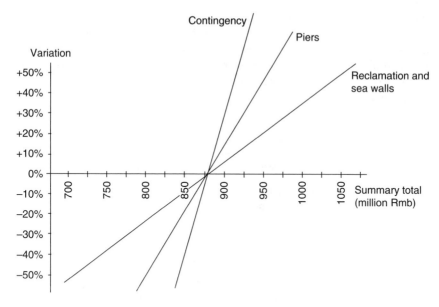

Figure 3.17 Sensitivity analysis – South China land reclamation.

3.8   A bridge over the Thames

In this section we explore the risk analysis process on a new bridge. We have been retained as project managers for the preliminary appraisal and design phases of the bridge and we are required to advise the government department responsible on the finance necessary to procure the project. The approximate estimates we received from the design engineers are shown in Table 3.19.

Table 3.19 Engineers' estimates at July 2004 in current prices (£ millions)

Contractor's design fee	2.50
Piling	25.50
Pile caps	2.75
Abutments	2.50
Precast beams and columns	50.00
Precast deck	10.00
In situ concrete works	1.10
Pavements	5.00
Utilities and markings	1.00
Preliminaries	8.00
London Borough of Greenwich additional requirements	
Access restrictions, noise restrictions etc.	2.50
Uncharted utilities, diversions etc.	2.00

We are specifically required to provide the treasury with a probabilistic forecast of project cost. We spend an intensive two days with the team of design engineers, the result of which is a set of probabilistic forecasts for the various parts of the project shown in Table 3.20. Once a Monte-Carlo simulation has been completed, the summary of the results can be tabulated in Table 3.21.

The probability density distribution and cumulative density distribution histograms are shown in Figures 3.18 and 3.19.

Table 3.20 Data input – project: Greenwich Bridge

Item	Description	Distribution	Lower	1st best	2nd best	Upper	Operator
1	Contractor's design fee	Rectangular	2.50			3.00	+
2	Piling	Trapezoidal	25.00	30.00	32.00	38.00	+
3	Pile caps	Triangular	2.50	2.75		3.25	+
4	Abutments	Triangular	2.50	2.80		4.50	+
5	Precast beams/columns	Triangular	45.00	50.00		60.00	+
6	Precast deck	Trapezoidal	8.50	9.25	10.00	11.50	+
7	In situ concrete work	Triangular	0.80	1.10		1.80	+
8	Pavements	Triangular	4.20	5.00		5.80	+
9	Utilities and markings	Rectangular	1.50			2.00	+
10	Preliminaries	Triangular	8.00	10.00		11.50	+
11	Additional requirements	Triangular	0.50	2.50		3.50	+
12	Uncharted utilities	Triangular	0.25	2.00		2.75	+
13	Fluctuation	Triangular	1.04	1.07		1.12	*

Table 3.21 Summary of Monte Carlo simulation results

Band	Band range	Frequency	Cum. frequency	Band	Band range	Frequency	Cum. frequency
1	105.30–107.30	0	0	16	135.31–137.31	97	871
2	107.30–109.30	0	0	17	137.31–139.31	65	936
3	109.30–111.30	0	0	18	139.31–141.31	44	980
4	111.30–113.30	0	0	19	141.31–143.31	10	990
5	113.30–115.30	0	0	20	143.31–145.31	8	998
6	115.30–117.30	0	0	21	145.31–147.30	2	1000
7	117.30–119.30	0	0	22	147.31–149.31	0	1000
8	119.30–121.30	0	0	23	149.31–151.31	0	1000
9	121.30–123.30	13	13	24	151.31–153.31	0	1000
10	123.30–125.30	38	51	25	153.31–155.31	0	1000
11	125.30–127.30	83	134	26	155.31–157.31	0	1000
12	127.30–129.30	131	265	27	157.31–159.31	0	1000
13	129.30–131.31	170	435	28	159.31–161.31	0	1000
14	131.31–133.31	184	619	29	161.31–163.31	0	1000
15	133.31–135.31	155	774	30	163.31–165.31	0	1000

Range:	121.52–147.06	Standard deviation:	4.34	3rd quartile: 134.31
Mean:	132.17	Median:	132.31	
1st quartile:	128.30			

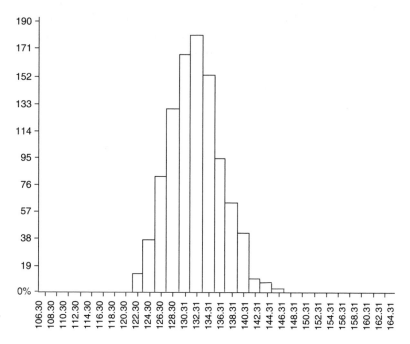

Figure 3.18 Probability distribution.

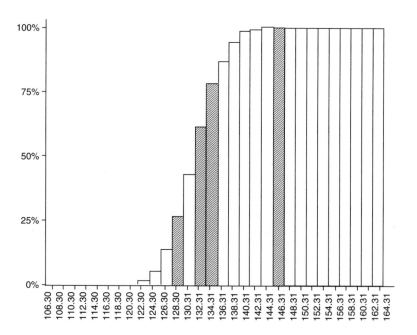

Figure 3.19 Cumulative probability distribution.

We can now tell the financier that the estimate of construction costs for the bridge will be £132 million and that the standard deviation (possible variance in project costs) is £4.34 million. Additionally, we can report that there is a 75 per cent chance of getting the bridge at the cost of £134.31 million or less.

Chapter 4

Perceptions of risk

4.1 Introduction

In Chapter 1 we showed that people's subjective perceptions of risk often differ from the objective assessments made by experts and scientists. We argued that managers must interpret risks in terms of human values and emotions and that traditional approaches to risk management have failed to take into account. This is one of the reasons why an increasing number of projects are adversely affected by seemingly irrational public responses and interventions. Although experts may claim that their objective calculations of risk are more accurate than a layperson's perceptions of risk, to the people that hold these perceptions, it is the objective assessments that are irrational. Therefore, public perceptions of risk cannot be discounted on the basis that they are uninformed, irrational, biased and subject to error. As the UK Engineering Council warned in its Guidelines on Risk Issues, "Engineering risk assessment is based on objective consideration of likelihood and consequences. However, individuals and often organisations have to make judgements based on their perceptions of likelihood and consequences. Thus the absence of a common framework for evaluating risk can make it difficult to arrive at consensus decisions" (EC 1999: 4). It is clear that different risk perceptions among project stakeholders introduce scope for confusion in communicating about risk and that great care needs to be taken in avoiding misunderstanding. Thus, all professions involved in engineering and building projects and in managing risk should understand the factors that can cause this to happen. They should also appreciate the levels of risk which society considers acceptable and tolerable. Ultimately, this is how the quality of their work will be judged.

In this section we seek to explain why people may have different perceptions of risk. In particular, we will explore patterns of bias in people's perceptions of risk and ask whether it is possible for a trained professional to anticipate and pre-emptively adjust for this. This will help decision-makers better manage responses to their projects by accommodating those perceptions in the risk management process.

4.2 The psychology on risk

The idea that risks are mediated through human processes has attracted a variety of explanations for over three decades and there is now a very large and complex psychological literature on the subject of risk perceptions and judgemental bias. Although much of the early work was disputed, mostly for being carried out in laboratory settings rather than in the real world, there is now a large body of agreement on how people perceive and assess risks and why they do so in different ways. Initially, the *economic perspective* assumed that people respond rationally to risks on the basis of comparisons of costs and benefits, reacting best to the path of maximum potential benefit (utility) to them personally. Subsequently, this idea has been developed and somewhat discredited by the *psychological perspective* which has found that individuals do not always behave rationally by basing their risk judgements on expected values. Rather, it seems as if there are a whole range of personal biases, shaped by past experiences, culture and education, that can cause people to attenuate (play down) or amplify (play up) a risk (Tversky and Kahneman 1981). More recently, the *sociological perspective* has found that reactions to risks are also the result of social processes (Barnes 2002). In other words, people do not react to risks in response to physical impacts alone but in terms of what it means to the community, physical or virtual, in which they are imbedded. This reliance on community as a source of perceptions is particularly strong when there is mistrust in external regulators or poor information about a risk. This has important implications for managers dealing with large projects that have a significant impact upon the community. First, it makes it clear that one cannot understand or manage the reaction to a decision without understanding how a community functions. Second, it emphasises the critical importance of effective communication to the risk management process. Kasperson and Kasperson (1996) have found that particularly important communication factors in shaping perceptions are the extent of media coverage, the volume of information provided, the way in which risk is framed and the symbols, metaphors and discourse used in depicting and characterising a risk.

4.3 Personal and reporting bias

The two main types of bias to emerge from the above discussion are *personal biases* associated with people's own psychological make-up, education, experience, culture, beliefs, values etc. and *reporting biases* which are introduced when people talk or report to each other. We do not propose that these considerations should be explicitly introduced into calculations of risk. However, they should be considered when developing a stakeholder consultation strategy. Ultimately, if a risk manager is to address risk as perceived by the public, then he or she must discover what those perceptions are and the various reasons why they might be different to his or her own. There is little point undertaking a rigorous risk assessment if those who could be affected by it cannot understand it. Furthermore, an understanding of the different risk perceptions held by different stakeholders allow the development of strategies to balance them. Different risk perceptions among project stakeholders are a major cause of conflict in projects and it may be possible to eliminate them through education and training, effective consultation, participation in decision-making or simply presenting risks in a particular way.

4.3.1   Personal bias

Personal bias in risk assessment can arise in a number of ways. In this section we describe common sources of personal bias and potential strategies for minimising it in a decision-making context.

4.3.1.1   Rules of thumb

It is known that people use rules of thumb (sometimes called heuristics) in decision-making. This is a rational response to a complex problem, especially where there is not enough relevant information available or where there is inadequate time to gather it. A good example of a rule of thumb is the 80/20 (Pareto) rule which implies that 80 per cent of the costs of a project are contained within 20 per cent of the units of finished work. Clearly, it is unlikely that the ratio will be exactly 80/20 in any one project and even less likely that it will remain so, from one project to the next. Whether 80 per cent of the costs are covered by 20 per cent of items depends on the nature of the project, the method of measurement used and who did the pricing. However, it might be true to say that "there is a propensity for a large proportion of the cost of any given project to be accounted for by a relatively small number of expensive items. Consequently, on many projects there is likely to be a large number of relatively cost insignificant items". This unfortunately does not trip off the tongue as easily as 80/20.

It is important to remember that rules of thumb are just that. They are usually anecdotal and should not be followed precisely. The main problem with rules of thumb is that they are usually based on previous experience, which means that when underlying conditions change, the rule is no longer appropriate. While they can be useful in some situations, they should generally be avoided and certainly should never be relied upon to make serious decisions.

Other rules of thumb which are commonly used by people when making decisions are the *representativeness* heuristic, *availability* heuristic and *adjustment* heuristic. The practical implications of each are discussed below.

REPRESENTATIVENESS HEURISTIC

This heuristic leads to fallacies about base rates and sample size and refers to people's tendency to ignore population base rates in favour of diagnostic information derived from small samples. It is related to the law of large numbers, which states that sample means tend to become closer to the population mean and less variable as the sample becomes bigger. Thus, the mean from a large sample tends to be more stable than that from a small sample and the sample variance is comparatively smaller. Neglecting this law of large numbers in making an estimate of risk is said to be committing a base rate or sample size fallacy, as is illustrated in the following examples.

1   The history of sub-contractor A is 25 per cent of all projects completed on time. You have examined the last three contracts performed by this contractor and all have been completed on time. What is the chance that the forthcoming project, if awarded to contractor A, will be completed on time?

    A   50 per cent
    B   25 per cent
    C   75 per cent.

The base rate suggests that B is the correct answer, i.e. there is a 0.25 likelihood that the contract will be completed on time. However, there is often a tendency for people to commit the base rate fallacy and to ignore this base rate in favour of the diagnostic information derived from the last three projects. This results in people returning answers A or C. In other words, they will assume that on the basis of the last three contracts, the contractor's record has improved. Psychologists would argue that these three readings are not enough to negate the base rate. However, this argument is based on the assumption that the number of readings used to form the base rate was large in comparison to the number of readings in the diagnostic information. The larger the number of readings forming the base rate, the less valid a judgement based on three recent projects. The second assumption is the one beloved to all economists, namely "that other things are equal". In this case, the base rate fallacy will have occurred only if nothing has changed in the contractor's behaviour, systems, management, personnel etc. to warrant a permanent change in performance.

2   A contractor has been served by two steelwork sub-contractors for a long period. The larger sub-contractor completes about 45 projects per year and the smaller sub-contractor completes about 15 projects per year. Overall the contractor's experience over 10 years of using these sub-contractors is that the overall delay on steelwork projects is 10 per cent. However, the contractor has never analysed the success of individual

sub-contractors and, thus, cannot accurately judge which sub-contractor is most reliable.

The contractor is introducing a new risk management system which requires it to collect more accurate performance data. To this end, it asks each sub-contractor to collect performance data over a period of 1 year. The contractor is particularly interested in knowing the number of deviations by each sub-contractor from the average 10 per cent success rate. Which sub-contractor will record the greatest number of deviations from the 10 per cent average success rate?

A   The larger sub-contractor
B   The smaller sub-contractor
C   Number of deviations from both sub-contractors will be the same.

To repeat, the law of large numbers states that the mean calculated from a sample will become closer to the underlying population mean as the sample gets larger. Means calculated from small samples are less stable than means based on large samples. Logically, therefore, the correct answer is B. Namely, the smaller sub-contractor will be more likely to submit the greater number of variant readings, where its projects are deviating more than 10 per cent over time. This disadvantages the small sub-contractor. However, over a large number of experiments, psychologists have found that people tend to return an answer of A or C by ignoring the law of large numbers.

Understanding the base rate and sample size fallacies is important since it forces decision-makers to ask fundamental questions about whether underlying conditions have changed before making any forecasts on judgement and to question whether diagnostic information makes a sufficiently strong case for adjusting longer-term base rate trends.

AVAILABILITY HEURISTIC

This is a heuristic which makes use of the retrievability of information and instances. Some event with which one has had personal contact looms more likely than it would statistically. Some information which is more easily available is more likely to be used as the basis for a judgement. For example, if someone has had personal contact with an aeroplane accident, a relatively rare event, then we are highly likely to overestimate the likelihood of future similar mishaps. The following example illustrates the danger of the availability heuristic.

You are an estimator employed by ACME contractors Ltd. You are asked to prepare a lump sum estimate for the pile caps. This estimate is to be used to compare the tender submissions of several sub-contractors.

The pile cap contract is to be let as a separate sub-contract to one of the sub-contractors.

You have checked a previous BQ from your company's records and found the following rates.

1   Reinforced concrete, $25\,N/mm^2$ filled into formwork – $90.00/m^3$.
2   High tensile reinforcement bars, bent and bundled, 16–32 mm diameter – $1050/tonne.
3   Formwork to foundation and beds, height $>1\,m$ – $60/m^2$.

There is significant evidence to indicate that people would be likely to use this information as the basis of an estimate without fully questioning the basis of the original estimates compared to those of the current estimate or whether these rates turned out to be an accurate reflection of market rates. One of the reasons for this is that this information is often not available. For example, previous estimators may not have properly recorded the assumptions on which they based their estimates and feedback information about the accurately of previous BQ (Bill of Quantities) rates may not have been collected at the end of the contract.

ADJUSTMENT HEURISTIC

This refers to the cases when people make estimates by starting from an initial value (an anchor) and adjust it to yield a final answer. The final result is usually biased towards the initial value. Anchoring occurs not only when a starting point is given, but also when an estimator has to base his or her information on the results of incomplete computation. The following example illustrates the anchoring process.

A client approaches you and wants to build a new supermarket. You have to prepare an estimate for the construction costs for this project. Your client tells you that he has learnt that your company has recently finished a similar supermarket for one of his competitors for $4.5 million.

The anchoring process would involve you making the mistake of using the $4.5 million figure as a basis of your estimate and trying to adjusting it for the circumstances of your project. Psychologists have found that the results would be an estimate biased towards the $4.5 million figure.

4.3.1.2   Question framing

Tversky and Koehler (1994) showed that the way people perceive risks depends critically on the way a question is framed or posed. A risk might appear to be acceptable when looked at in aggregate but can become unacceptable when broken down into its constituent parts. For example, consider the construction of a road project through a relatively unknown yet unspoilt forest in Tasmania, Australia. To most people, the risk that this project will affect their lives adversely will seem quite low. However, if they are presented with a long list of how it could lead to the felling of ancient rainforest, the destruction of important habitat, a greater chance of flooding and landslides for local communities etc. the risks suddenly appear to be a lot larger.

While one would expect laypeople to be most vulnerable to such biases, Morgan and Keith (1995) illustrated that experts are just as susceptible. They assembled a group of climate experts to discuss the probability that temperatures would rise if CO_2 omissions doubled. After being asked for an expert assessment, the experts were asked to put together a research programme for investigating this effect in detail. Although this gave them no extra information to estimate the probability that temperatures would rise, when they were asked again for their assessment, their estimates rose significantly. Of course, it was only the perception that had changed, not the risk itself.

Good negotiators use this phenomenon to their advantage by framing questions in a way which reinforces their arguments. For example, by with-holding detailed information, negotiators can encourage opponents to underestimate the risks in a solution they propose. Skilled lawyers also use this tactic to manipulate perceptions of guilt in juries. For example, if a lawyer wants to create a feeling of uncertainty and discredit someone, they ask the same question numerous times, increasing the quantity and complexity of facts to the witness. In contrast, if they want definitive statements from a witness which provide confidence for juries, then the questions put will not be encumbered with too many facts or details.

4.3.1.3   Emotions

The emotional detachment of experts and scientists from the world they seek to measure is one reason why differences often exist between public and scient-ific perceptions of risk (Slovic *et al.* 1981). For example, construction safety risks are often officially measured in terms of fatalities and serious incidents. However, employee perceptions might be more concerned with the longer-term suffering caused by accidents, considering the possibility that being maimed for life may be a fate worse than death. To other people, safety risks may be judged more on ethical grounds and violations of human rights rather than on loss of life or longer-term suffering. Others might be more concerned with the psychological aspects of safety such as stress, burnout and damage to family life. These emotive dimensions of safety risks do not lend themselves to accurate modelling or calculation and therefore tend to be ignored by experts, scientists and decision-makers. Furthermore, there is often a fear among decision-makers and scientists that if these emotional dimensions are revealed, stakeholders will react irrationally. Despite these fears, it is important not to ignore these aspects of risk. The consequences of doing so can be disastrous as has been illustrated by a number of high-profile public health crises in recent years. Of particular note were the BSE crisis in the UK and the SARS and bird flue epidemics in Asia, where a whole series of politically motivated decisions resulted in a denial in the early stages, that these diseases might be transferable to humans, in fear that the public would overreact (Klein 2000). The consequences of these cover-ups were severe, involving massive damage to associated industries and a crisis in public confidence in government pronouncements on risks in other areas, such as the introduction of genetically modified food. These lessons are relevant to industries like construction which are experiencing increasingly emotional interactions from the public over safety and environmental issues.

4.3.1.4   Familiarity and control

Another factor that has been found to influence perceptions of risk is familiarity with the hazard. There is evidence that people who have become accustomed to living in risky environments naturally underestimate the risks to their safety. Psychologists call this process habituation and it occurs when repeated stimulation by a risk reduces the fear associated with it. For example, risks such as smoking and drinking are tolerated because they are familiar, whereas new risks such as genetically modified food are not. Another good example can be found in people living near nuclear power stations who have a lower perception of nuclear risk than those who live further away. Their repeated exposure to the fear-stimulus coupled with the lack of catastrophe, gradually weakens the power of the stimulus to evoke a fear response. Furthermore, people in situations such as this tend to develop coping/survival mechanisms to reduce the impact of the risk. These responses may be psychological such as "denial" or more tangible such as putting a nuclear bunker in their garden or having a special escape plan.

The implications of this habituation tendency are profound for decision-makers involved in high-risk projects because it indicates that, as time goes by, people become desensitised to the risks that surround them and need constant reminding to keep them alert. Furthermore, it indicates that people within a project are more likely to ignore risks than do outsiders. Yet paradoxically, the normal perception is that outsiders are misinformed (Turner and Pigeon 1997). This suggests that project risks may be better analysed by external people who are less likely to be affected by habituation.

4.3.1.5   Responsibility

One of the most important influences upon people's perceptions of risk is their perception of responsibility for it. People tend to be more sensitive to the risks that they are responsible for. Indeed, confused responsibilities for the monitoring of different risks have been the reason for the downfall of many organisations. For example, the Bearings Bank crisis in the mid-1990s was caused by muddled lines of responsibility and accountability for the reporting of potentially damaging events (Sheaffer *et al.* 1998). This confusion, coupled with huge time pressures under which people worked, allowed Traders legitimately to override the controls that were in place without any detection. Similar problems exist within construction projects where it has been known for some time that one of the major problems has been confused responsibilities for risks caused by complex, voluminous and legalistic contracts (Barnes 1991). The result has been a widespread failure to manage risks effectively, thereby causing substantial time and cost overruns to become common on many projects.

4.3.1.6 Evolution

Perceptions of risk are shaped by the way in which a risk evolves. In particular, those risks that emerge *gradually* tend to be underestimated compared to those that emerge *suddenly*. For example, this is one of the reasons why most smokers underestimate the risks of their habit (Hersch 1998). For project managers, this phenomenon has important implications. For example, in the construction industry there is evidence that as many as 60 per cent of the problems that occur during projects are built into the project during the design stage and that they accumulate slowly, largely unnoticed over time, as they are hidden by subsequent design decisions. The problem is that when they do surface, their neglect has allowed them to escalate into much larger problems which are much more difficult to resolve. Of course, the opposite is true in the case of potential opportunities.

4.3.1.7 Invincibility

People have a natural tendency to view themselves as immune from some risks, seeing others as the primary candidates for accidents. For example, most people view themselves as better drivers than they really are. Day after day their experiences of driving over the speed limit without an accident confirm this. Indeed, when they do hear of accidents it is usually through the media, which confirms that accidents only happen to someone else. While this example might seem irrelevant to project management, it is likely that people could develop similar perceptions about certain work-related risks. Therefore, it is a manager's responsibility to dispel this sense of invincibility by continually reminding decision-makers about everyone's vulnerability to risk. Interestingly, this applies as much, if not more, to experts who are often less aware of their own limitations than laypeople. This was vividly illustrated in the Teton Dam collapse in America in 1976 where the official report attributed the disaster to the unwarranted overconfidence of engineers who were absolutely certain that they had solved the many serious problems that arose during construction (US Government 1976).

4.3.1.8   Unpleasantness

People have a natural tendency to bury unpleasant memories in their sub-conscious, a process which is known as *repression* in psychology. This causes people to underestimate risks which are perceived to be particularly disagreeable. Repression is an important process in organisations, particularly in relation to health and safety risks, which tend to be most closely associated with physical injury and pain. This can cause such risks to be underestimated.

Repression is a phenomenon which requires shock tactics to reverse. This is what justifies the increasingly vivid and controversial road safety and anti-smoking advertisements seen on TV which are designed to shock people into releasing their suppressed memories. Although similar tactics could also be used in the management of safety risks on projects, care should be taken. Not only are there important ethical questions to be answered but there is still considerable controversy surrounding the point at which a shocking image drives perceptions of risk more deeply into a person's subconscious.

4.3.1.9   Preparedness

Some risks require managers to do things they prefer not to do or that they are not prepared to deal with. Argyris (1990) found that such risks are likely to be played down or overlooked because those who reveal them are likely to be castigated. At the very least, these people are required to provide higher standards of evidence than those who provide information that supports existing expectations and hypotheses. The potential dangers of this defensive behaviour have been vividly illustrated in many disasters. For example, the Hillsborough football stadium disaster in the UK and the Challenger space shuttle disaster in the US were caused by decision-makers filtering out the advance signs of impending disaster which they were not prepared to deal with (Jarman and Kouzmin 1990). Managers can avoid this problem by regularly changing project team membership, however inconvenient this may be.

4.3.1.10   Priorities

Most projects have competing goals, and perceptions of risk are strongly influenced by the relative priorities that are put on them. The new extension to London's underground Jubilee Line was a good illustration of how risks in one area can be overlooked when priorities lie elsewhere. In this project, managers traded-off risks in areas such as safety against those which affected the perceived priorities of controlling time and costs (Glackin and Barrie 1998). The result was a poor safety record and a number of acrimonious and highly publicised industrial relations disputes. Another example of how competing priorities influence perceptions of risk can be found in the smoking habits of teenagers. It has been found that they tend to view smoking as a lower risk than do older people because it serves a strong social function and is also a demonstration of defiance to parental norms (Hersch 1998).

4.3.1.11   Blame

Perceptions of blame and the use of negative reinforcement are strongly linked to perceptions of risk. For example, Ryan and Oestreich (1998) found that in organisations where blame and recrimination is common, people tend to suppress and underestimate the potential risks they see in fear of being blamed. In contrast, open, trusting and participative environments, which emphasise positive reinforcement, engender a sense of collective responsibility for monitoring risks, which means that fewer are overlooked. This has important implications for many projects where excessive competitive tendering, onerous contractual practices and fragmented organisational structures and supply chains make it difficult to achieve any sense of openness and collective responsibility for the management of risks (Latham 1994).

4.3.1.12   Information technology

Dramatic improvements in IT in recent years have altered our perceptions of risk. For example, some suggest that IT is responsible for permanently lowering equity premiums on US money markets by increasing the availability of real-time information which is capable of being analysed (Greenspan 1999). IT permits the continuous monitoring and analysis of risks and, because of this increased access to information, there is little doubt that those who invest in IT will have vastly different perceptions of risk than those who do not. However, it is a mistake to think that investing in a sophisticated IT system will reduce risk. While IT can increase the flow of information in organisations, it can often reduce the quality of communication. For example, many organisations find that people use email to communicate when they may have previously talked to each other. In many instances, this can be a far less effective means of communication. Furthermore, with the growth of email and web transactions, information overload has become a major issue, as have information security and the theft of intellectual property (Fenton-Jones 2003). It appears that IT presents companies with as many risks as opportunities and should be seen as a mechanism to improve the exchange of knowledge and ideas within a firm, not as a substitute for it.

4.3.1.13 Compensation and irreversibility

People need to feel that they will be protected from risks by promises of compensation or reversibility. When this type of protection is afforded, risk tolerance tends to increase, resulting in an attenuation effect. In contrast, the more a danger is perceived to be irreversible or non-compensated, the more it is resisted and amplified. This is relevant in industries such as construction, due to the widespread use of lump sum and onerous contracts, which often offer inadequate compensation for those bearing risks. All too often, risks are passed down the contractual chain, using back-to-back contracts, to the point of least resistance where there are inadequate resources, expertise and contractual provisions to deal with risks that eventuate. While, in theory, bids can be loaded to reflect risks, excessive price competition often prevents this from happening. Employers are then left with the illusion that they have successfully transferred a risk to another party whereas in reality, when that risks arises, that party will do everything in their power to avoid it, resulting in greater long term and often hidden risks. A well-known principle of good risk management is the importance of appropriate compensation for the transfer of risk. If risks are not properly compensated, then clients will pay in less obvious and often more painful ways.

4.3.1.14 Seriousness

Research has shown that in assessing a risk, the public's perception of impact holds more weight than the *probability* of it happening. The reason for this is related to the difficulty most people have in understanding expressions of probability compared to those of consequences. This is undoubtedly one of the major factors in the public's perception of risk from nuclear power where the risks may be extremely low but the consequences of an accident are catastrophic. The critical factor in considering consequences seems to be its magnitude rather than its frequency. For example, the public perception after September 11, 2001, that working in tall buildings was unsafe was formed more by vivid images of the tragedy that killed thousands of people in one instant than by the fact that this event has only ever occurred once. This phenomenon is particularly relevant to high-risk industries like construction where a few high-impact disasters can tarnish the whole industry's image, which overall may be making good progress towards reducing adverse impacts. The phenomenon also works in the other direction. For example, because construction tends to suffer a relatively large number of minor unreported accidents, perceptions of risk could become attenuated, creating the possibility that many will be overlooked. Educational programmes that provide comprehensive and accurate information to decision-makers and stakeholders can help to alleviate this problem.

4.3.1.15 Presentation

Perceptions of risk depend on how data relating to risk are presented. For example, if someone is informed that the chances of having a fatal car accident is only about 1 in every 3,500,000 trips, then refusing to wear a seat belt might seem quite reasonable. However, when one is asked to consider their lifetime of trips over 50 years of driving (say, 40,000 trips) the probability of being killed rises to 0.01 (1 in 100) which portrays a profoundly different perception of risk. This indicates that in creating an educational programme about risks, every decision about the format and presentation of information is important, whether it be via training courses, labels on materials, media presentations or leaflets. It is also important to realise that some media of communication can have a different impact than others. For example, it might be the case that certain groups of people do not read certain publications or do not have access to certain media such as email or the World Wide Web.

4.3.1.16 Culture

Most projects are served by a number of professions, all having different perceptions of where the boundaries of each other's risks lie. For example, in the construction industry, Ibbs and Ashley (1987) found that a building contractor was almost twice as likely to believe that the risk of sub-surface investigation had been allocated to the owner than the owner was himself. According to Taylor (2000), these differences in risk perceptions are due to a number of factors, including the traditional roles played on projects by different professionals, the increasing encroachment of professions on each other's territories, the inability of professional institutions to precisely define the scope of their activities in their contracts and unrecorded changes to services being made during a project. However, a more fundamental cause of these boundary disputes is the different cultures which characterise each profession, causing them to see and interpret the world in a different way. For example, Lawson (1979) found that architects and engineers solve problems in different ways and, consequently, would probably develop different perceptions of the risks associated with a task. For example, an engineer would probably be far more sensitive to the construction risks associated with a particular design detail than an architect who might be more concerned with the functional risks. Furthermore, different occupations tend to attract people with specific bio-social and psychological attributes and this can also cause risks to be perceived in a particular way (Pierre *et al.* 1996).

While it is difficult for managers to change professional cultures, they can reduce the barriers between them by using multi-disciplinary teams or by choosing procurement methods which break down disciplinary boundaries such as design and construct and partnering.

4.3.1.17   Trust

The more that consultation is seen as a token exercise, the more a risk will be amplified and resisted. In contrast, the more that stakeholders feel that their opinions have been taken into account in the making of a decision, the more tolerant they are of risks associated with it. This is because meaningful consultation, particularly early in the decision-making process, develops trust in the decision-maker. Nevertheless, trust is a fragile emotion which is created slowly but destroyed quickly. This means that decision-makers must work hard and continuously to ensure that their consultations are seen as a genuine reflection of mutual concern for the interests of all affected by a decision.

4.3.2   Reporting bias

We have pointed out that there are fertile conditions for error when an estimate/forecast is produced by an expert or consultant and is then reported to a director, partner or some other senior decision-maker. Here we explore two important sources of reporting bias, namely the method of managerial control (positive or negative reinforcement) for over- and under-estimation and the bias and misinterpretations which result from the differences in personal risk attitudes.

4.3.2.1   Methods of management control

In most organisations there are systems which require managers to report back with explanations when projects exceed budgets or when returns fail to meet the initial target by some arbitrary cut-off figure (often 10 per cent). In contrast, where a project comes in under budget, or when returns exceed forecasts by more than the same amount, the explanations, although necessary, are not examined in quite the same detail.

When an organisation applies negative sanctions to managers who exceed budgets and does not praise or reward when projects come in on, or under, budgets, there will be a natural tendency to be risk averse – overestimating forecasts of project costs and underestimating forecasts of project revenues. The likely consequence will be that viable projects will not get supported and that resources will be misallocated.

Figures 4.1 and 4.2 illustrate, respectively, internal rate of return (IRR) and project cost forecasts, showing in each case the areas within which conservative and risk seeking forecasts lie. A conservative and self-preserving estimate of project *costs* will lie in the upper range of possible outcomes. We may regard as risk seeking any estimate from the lower end of the range, with a probability of being exceeded of at least 0.5. Conversely, when we consider forecasts of project *returns*, the conservative forecaster will draw a forecast from the lower end of the range. This is

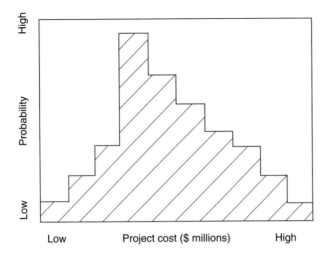

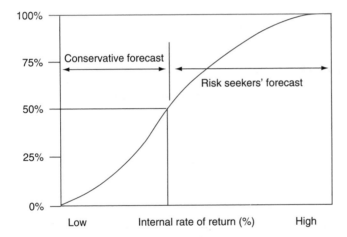

Figure 4.1 Risk attitude in forecast of returns.

a very safe forecast since it will be very unlikely that the project will disappoint its sponsors by producing no return. Our choice of 0.5 as the cut-off point is purely arbitrary for the purposes of this discussion. In reality, an organisation may choose a different cut-off, dependent on its own corporate risk attitude. A risk-taking organisation would set a lower cost cut-off (making lower cost estimates appear less risky) and a higher IRR cut-off (making higher IRR estimates seem less risky) and vice versa for a risk averse organisation.

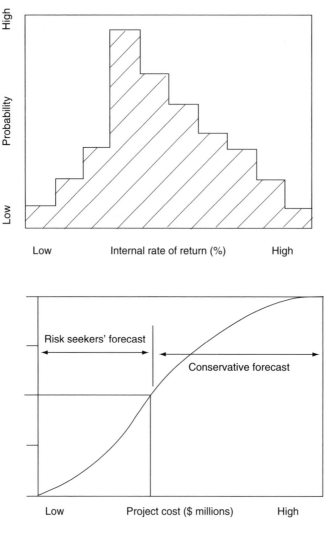

Figure 4.2 Risk attitude in forecast of costs.

4.3.2.2   Differences in personal risk attitude

Biases and misinterpretations may occur even in situations where the output from a cost model is reported in ways which purport to take account of risk exposure. For example, a consultant may indicate that there is a "good" chance that the completed project in a specific location will achieve a rent of say $70 per square foot (knowing that when the project is finished, the state of the property market could be the same as, better or worse than today). Alternatively, an advisor may state that there is a "reasonable" chance that a project can be completed for less than \$40 million. What do these statements actually mean? The language in itself seems reasonably clear. However, is a "good" chance 9 in 10, 8 in 10 or 6 in 10? Is a "reasonable" chance 8 in 10, 7 in 10 or 5 in 10? These differences could be very significant to a decision-maker deciding between different options. What does the decision-maker to do? It would seem rational to try to deduce the influences acting upon the forecasters (such as the reinforcement system as discussed above) and make adjustments for this before making a decision. If you think there are influences making them conservative then you increase the projected project return by, say, 10 per cent. Such adjustments are of course completely arbitrary and can lead to equally inappropriate decisions. Indeed the same sort of argument could be applied to multiple single point forecasts such as those employing the three points – most likely, optimistic and pessimistic. The precise point at which the optimistic and pessimistic values are located is a function of the risk attitude of the forecaster. This is unlikely to be identical to the risk attitude of the decision-maker and therefore may lead to misinterpretations of forecasts. For example, what is a pessimistic forecast to a forecaster might be an optimistic forecast to a decision-maker. In other words, the decision-maker may be far more risk averse than the forecaster. However, even if the decision-maker decides to adjust for the risk attitude of the forecaster by reducing forecasts of returns and increasing forecasts of costs, the nature of the adjustment is likely to be quite arbitrary and thus may also result in bias. These problems are well illustrated in the case study at the end of this chapter.

4.4   Eliminating personal and reporting bias

It is impossible to eliminate reporting biases which may result from differences in risk attitudes between people in reporting chains. It is also very difficult to disconnect risk management performance and reporting from pay and promotional structures. In reality, the best that can be expected is to require decision-makers to make as explicit as possible any assumptions made in making a decision. It is also important to ensure that decision-makers are familiar with notions of personal and reporting bias and to introduce procedures which incorporate feedback loops into decision-making processes. Finally, risk analysis should as far as possible be undertaken by independent experts and separated from the decision of how to respond to that risk. This separation is critical so that decision-makers do not directly influence a risk assessment based on their prejudices or external stakeholder pressures on how the risk should be managed. It is this argument which is used by risk analysts to defend their independence and to distinguish between good decisions and popular decisions. For example, a risk analysis may indicate a decision which is very unpopular with some stakeholders but which will produce the best outcomes for a company. Analysts argue that it is crucial that risk assessments are made free from bias and that they are based as much on objective facts as possible. It is then up to the decision-maker to take societal values into account – not the assessor. However, while the complete separation of those responsible for risk assessment and decision-making is ideal, in reality there are many reasons why this is unrealistic. For example, it is highly likely that a risk assessor will know of the pressures that a decision-maker is under in making a decision. There is therefore a real chance that they too will be influenced by them, even if there is a conscious effort not to be. Furthermore, a good assessor should have the knowledge to indicate how intervention can best reduce the risk and it would be futile to waste this expert knowledge.

4.5   Conclusion

In this chapter we have discussed the psychology of risk in practical terms. We have argued that the manager who relies wholly on scientific expertise and ignores the human dimension of risk assessment is likely to create more risks than he or she solves, even in the most technical situations. While our technical skills have come far, the new challenge for managers need to rise above the limitations of individual minds, reconciling the interests of different stakeholders to reach a consensus about the risks which face a project and about the best way to deal with them.

We have shown that no matter how objective you try to be in measuring risks, it is inevitable that there will be some bias in the risk analysis process. We have argued that it is better to recognise and manage this than ignore it. In particular, we have shown how people make judgements about future events according to their own information, perceptions, beliefs and experiences. Not everyone can apply the same levels of skills, experience and resources that may be available to your organisation. Therefore, when you involve a variety of stakeholders, you will have a variety of different perceptions that are unlikely to correspond. To avoid potential conflict, these perceptions will have to be reconciled by ensuring that everyone has the same information about the risks concerned. Effective risk communication is a central and critical aspect of the risk management process.

We have also learnt that group dynamics can interfere with the risk analysis process. Therefore, it is wise to structure and manage risk assessment groups to ensure that views do not become polarised around strong or powerful individuals. You must also be wary of the potential influence of people's extended social networks. Although these may be difficult to penetrate, they exert a very powerful influence upon an individual's perceptions of risk and must also be managed. This is best done by gaining the trust of stakeholders. Finally, to minimise reporting bias, which can arise when information about a project passes from one person to another, it is good practice to minimise the number of people in the risk information supply chain.

At an individual level we also know that people tend to be inconsistent in identifying probabilities. This often arises from differences in access to information or between people's attitudes towards risk exposure. For example, risk-takers tend to make conservative estimates of risk whereas risk-averse people tend to do the opposite. Therefore, in assessing risks, it is good practice to try to understand your own and others' attitudes towards risk and account for it. You should also search for inconsistencies in individual perceptions of risk by asking the same question a number of different times and in different ways.

We have also shown that people tend to use heuristics to cope with the complexities and time-pressures of the real world. These are approximations to the truth and can be deceiving. Therefore, you should ensure that enough time and resources are dedicated to the risk assessment process and that people use reliable data wherever possible. This will also help to avoid the tendency for people to give too much weight to recent data. For example, if a sub-contractor normally completes contracts on time but in their last three contracts they have finished late, the tendency is to assume that the sub-contractor will be late on the next contract. It is bad practice to change base trends unless there is some strong evidence that alters the assumptions upon which the original estimate was based. For example, the sub-contractor may have had a change in management.

Finally, we have discovered that people tend to overestimate (amplify) or underestimate (attenuate) risks in different circumstances. Understanding and managing these circumstances is immensely important in predicting and managing *external* stakeholder responses to your project. It is also important to ensure that a balanced and reliable *internal* estimate of risk is made. For example, an amplification effect is likely to result in stronger external opposition or support for your project and possibly an over reaction to it. In contrast, attenuation could result in ambivalence. Internally, the amplification effect is likely to result in a tendency to play safe and avoid risks, while attenuation may result in dangerous risk taking behaviour. In particular, we have highlighted the factors that tend to lead to an amplification or attenuation effect. These are as follows:

People tend to amplify (overestimate) the likelihood of a future event when:

- There is a strong desire for something not to happen
- There is the possibility of punishment or recrimination for getting things wrong
- They are responsible for it
- They could be responsible for it or are unsure of who is responsible for it
- A risk is its constituent parts, rather than in aggregate
- A risk involves emotive and ethical issues
- There is no compensation if the risk eventuated
- A risk is a high priority in terms of achieving organisational objectives
- There are serious consequences if the risk occurred
- A risk depends upon a combination/series of events occurring (A safety-mechanism for overcoming the computational difficulties in arriving at a probability calculation)

- There is a lot of information about a risk (This means that knowledge can be a risk, which is why it is often said that people who make the most mistakes win tenders.)
- They have had recent experience of a risk. (For example, someone who has recently been involved in pollution incident is likely to overestimate the probability of a pollution incident when asked to estimate it.)

People tend to underestimate the likelihood of an event (take risks) when:

- There is a strong desire for something to happen
- There has been past success in controlling a risk (A sense of invincibility is one of the greatest sources of failure in contracts. It is often said that nothing fails like success.)
- A risk is familiar (There is previous experience in dealing with it.)
- There is more trust in a relationship which could cause a risk
- A risk is presented in aggregate rather than in its constituent parts
- The risk is a low priority, in terms of its impact on objectives
- A risk could be or has been unpleasant in its consequences
- It is assessed by an expert
- A risk emerges gradually
- There has been no preparation for the risk eventuating
- The risk is not emotive
- Compensation is available if a risk eventuated
- There is little information about a risk (ignorance is bliss)

We end this chapter with a case study which illustrates the practical application of the processes discussed above. It looks at the potential for reporting bias during the estimating process in a typical construction company when an estimator, who is preparing a bid, has to communicate with a managing director who will make the mark-up decision.

4.6   Saying what you think

In this case study we explore the potential for reporting bias when a contractor bid for a new commercial project. An estimator is just completing work on the bid and has to report a net cost estimate to the managing director who will make the mark-up decision. The estimating team gets together on the day before the bid is due and decide that their best estimate of the net cost is $72 million. This figure includes tangible and intangible costs, head office overheads and an allowance for the cost of recovering finance charges. It includes no profit and has been arrived at by breaking the project down as shown in Table 4.1.

What the manager thought the estimator said...
The manager receives this figure with the background briefing from the estimators and planning department. The mark-up decision is familiar to the manager

Table 4.1 Deterministic estimate of net project cost

Component	Cost ($ millions)
Materials	25
Labour	10
Sub-contractors	20
Overheads and finance	12
Supervision on site	5
Total	72

who is accustomed to taking calculated risks in order to secure work at favour-able rates for the firm. The manager knows that the estimate is a forecast of the outturn costs, should the firm win the project. Thus, it would be rational to assume that the estimate is the most likely figure drawn from a distribution. In estimates of costs, most managers are very conservative and expect a right skew. In other words, that there is more likelihood that the project will run over budget than under budget. The mental cost model of the project held in the manager's mind is illustrated in Figure 4.3. The model indicates that the most likely outturn cost is $72 million. The optimistic outcome is a project net cost of $65 million and the pessimistic outcome shows a cost of $86 million.

What the estimator thought but did not say...
In many of the risk analysis workshops undertaken it was possible to analyse, in detail, the deterministic project cost estimates. This exercise involved meeting the estimating teams towards the end of the bid period when they had a quite firm deterministic idea of the net cost of the project. In a two-hour session it was possible to carry out a detailed simulation of the project to produce a frequency histogram of possible outcomes and, from this, a cumulative probability graph. The experience of these analyses was that when the simulation results were obtained, it was found that the deterministic figure given earlier had been extremely conservative with a probability of

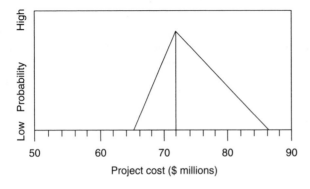

Figure 4.3 Managers perception of the single point estimate.

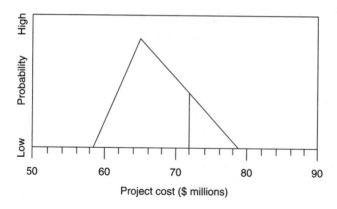

Figure 4.4 Estimator's perception of project cost.

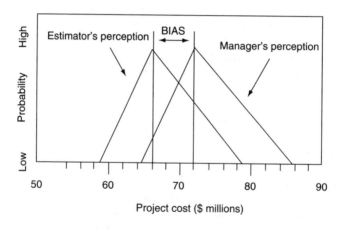

Figure 4.5 Forecasting bias.

final costs being below that estimated in the region of 0.9. This is not surprising given the long history of empirical studies demonstrating that most business people are markedly risk averse. Thus for the purposes of discussion, using that datum, we may construct what our estimator thought but did not say. This is illustrated in Figure 4.4 where the $72 million is a conservative estimate drawn from a distribution with lower and upper bounds of $59 and $79 million respectively. In this distribution, $72 million is not the expected figure but a highly conservative estimate, which is designed to give a large margin of safety to the estimator. The expected cost is in fact just under $66 million.

The amount of bias introduced is illustrated by overlaying Figures 4.3 and 4.4 and is depicted in Figure 4.5.

Chapter 5

Risk response, crisis management and recovery

5.1 Introduction

The decision about how to respond to risks and opportunities, having identified and analysed them, is the final stage of the risk management process. In essence, the decision is simple – *to do something* or *to do nothing*. While both decisions are rational responses in different situations, the aim should always be to establish a strategy to mitigate potential threats and maximise potential opportunities.

5.2 To do nothing

It is irrational to respond to every potential risk and opportunity an organisation faces. While some risks and opportunities may be so large that they justify an automatic response, almost independent of all other concerns, others are so small that they are not worth the trouble to think about. Such decisions rest on an analysis of the relative costs and benefits of responding within the context of time and resource constraints. For example, if the costs of responding to a risk or opportunity exceed the potential benefits of doing so, then a decision-maker may legitimately do nothing. Of course, the relative costs and benefits of different responses vary from company to company and inefficient companies with high costs should be careful not to ignore a problem on the basis of their inefficiency, rather than a reasonable estimate of costs. Furthermore, every company has a different appetite for risk that will dictate the relative perceived costs and benefits of different options. Therefore, understanding and clearly articulating that appetite is critical in formulating a response and ensuring it is consistently applied across a company. For example, some companies may adopt a *zero risk* policy in relation to certain risks, which means that *any* event with a potential impact would justify a response. While such a policy makes cost/benefit analysis irrelevant, in reality, most risks in most companies are managed by a policy which requires risks to be reduced to *as low as reasonably achievable*. In this context, what is considered a reasonable level of risk becomes critically important, not only in the eyes of the company but in the eyes of the law. It must always be remembered that the law courts are the final arbiter of whether a decision is appropriate or not. Indeed, since the law is applied and developed primarily through parliament and the jury process (often through consultation with community groups), it is in fact the public's perception of risk that is often a dominant factor in its decisions. This is yet another reason why the public perception of risk cannot be ignored by decision-makers.

In addition to situations where the costs of responding to a risk or opportunity exceed the benefits of doing so, a potential problem or opportunity can also be legitimately ignored when:

- Existing controls are adequate to maximise an opportunity or minimise a threat.
- The risk is low and within acceptable limits for a business.
- There is a high probability that a risk will disappear in the future.
- There is a high probability that an opportunity will increase in the future.
- There is a high probability that taking an opportunity will prevent others arising.
- There is a high probability that dealing with a problem will cause other problems to arise.

While doing nothing may be a rational response to a risk or opportunity in certain situations, it is very important not to confuse the idea of doing nothing with denying that a risk or opportunity exists. Denial is always indefensible and can result in disaster. Furthermore, doing nothing does not mean forgetting about a risk or opportunity. Once identified, any problem or opportunity must be continuously monitored to ensure that relative costs and benefits have not changed in a way which later justifies a response.

To many decision-makers, the above arguments will seem academic since the practicalities of not responding to a potential risk or opportunity are far more difficult than simply comparing costs and benefits. For example, precise information about potential costs and benefits is rarely available. Furthermore, there may be laws that demand a response or considerable social pressure from within and outside an organisation to do something about it. Finally, legal and moral judgements about the costs and benefits of response are often difficult, particularly in relation to issues such as health and safety where it would mean placing a value on someone's health. This uncomfortable ethical predicament is the economic dilemma of risk management in the area of health and safety and is probably one of the main reasons why people are reluctant to identify safety risks. Ethically, there is no such thing as a tolerable safety risk and once identified, some action will inevitably be required. Certainly, the person who identifies a safety risk and consciously decides to do nothing in response for economic reasons is likely to face unanimous condemnation if this results in a fatality or injury. As Horlick-Jones (1996) notes, in the aftermath of a disaster, the public and media have a tendency to embark on a process of ritual damnation turning on those who may have contributed to satisfy a desire for retribution. So it is a brave manager who consciously decides to do nothing in response to a risk. It is certainly not an easy option, particularly in the increasingly litigious business environment in which decision-makers operate. Consequently, most potential problems and opportunities will justify action of some kind, although the pressure to respond to opportunities is normally far less than the pressure to respond to potential problems.

5.3 To do something

When a potential opportunity justifies a response, the aim should be to maximise its potential benefits. In making this response, the choice is straightforward; to *wait in order to maximise the opportunity* or to *take the opportunity*. Opportunities should be *taken*, only when there is no further possibility of increasing their potential benefits to an organisation.

When a potential risk justifies a response, the aim should be to minimise its potential costs. In making this response there is a hierarchy of four choices starting with *avoidance* and moving to *elimination, reduction* and finally to the *transfer, sharing* or *acceptance* of any residual risks that remain. These choices are discussed below and it is likely that they will have to be combined in arriving at a complete and effective response to a specific risk.

5.3.1 Risk avoidance

The ideal option in responding to a threat is to *avoid it*. Unfortunately, this is not always possible or indeed desirable. For example, a company might decide to avoid the risk associated with a certain type of project by not tendering for them. However, this decision introduces other risks since work will have to be sought elsewhere. Furthermore, businesses make money by taking risks and managing them effectively, meaning that avoidance can result in missed opportunities as well as missed threats. It is crucial to remember that risk taking is the main source of wealth creation in business and to prosper; it is unavoidable that managers must take risks at some point. The trick is to avoid risks that cannot be controlled and to take and manage effectively risks that can be.

5.3.2 Risk reduction and elimination

In most situations, a more realistic option is to *reduce* the risk of threats and ideally *eliminate* them. This can be done in a number of ways, depending on the type of risk posed. For example, one could introduce training programmes or changes to working practices to prevent risk arising from human error. Alternatively, in the case of technical risks, one could change some aspect of a project's design, materials or technologies. In the case of political risks, lobbying and community consultation may be an effective risk reduction strategy to allay public misperceptions about a project. Finally, in reducing competitive risks another option might be the removal of a competitor by acquisition. Incidentally, acquisition may also be an option to reduce supply risks because by acquiring a key supplier one could ensure the supply of critical raw materials at a desired level of quality. Indeed, there are an infinite variety of strategies that can be used to reduce risk, which are limited only by the imagination of decision-makers. Of course, not everyone is capable of the same level of imagination and in the eyes of the law the most important thing is to justify that everything "reasonably practicable" has been done to mitigate a risk. Not doing so may result in court action as managers become increasingly accountable for their actions by governments, shareholders, market analysts, pressure groups and the general public.

Developing strategies to reduce risks requires knowledge of *causes*, *consequences*, *likelihood* and *controllability*. Clearly, when there are limited resources, it makes sense to target them on the causes of those risks which are most likely, are most controllable and have the greatest consequences if they occur. For example, on a fast-track project in central Sydney, local residents' concerns about noise and dust, which threatened to limit working hours on site, were appeased by involving them in project planning and by the provision of a $250,000 bond, which could be called on if work occurred outside agreed hours. This risk was highly likely with relatively controllable causes and the potential consequences of working hours being restricted was worth the substantial bond investment. In another project in the UK, operatives constructing caissons for a new bridge were exposed to contaminated river water which presented the risk of contracting hepatitis. After investigating the causes of this problem, the risks were reduced by changing working practices and by introducing a vaccination programme. However, as in most cases of risk mitigation, there was only so much the organisation could do to mitigate these risks within the timescale of a project. For instance, it was not possible to lobby government officials to investigate and tackle the cause of the pollution upstream. Like most project managers, decision-makers also had to decide how best to handle the *residual risks* which remained after all reasonably practicable risk reduction measures had been taken.

5.3.3   Dealing with residual risks

In dealing with residual risks which cannot be reduced or eliminated, the choice is simple once again to *retain them, transfer them* to some other party or *share them*. It is essential to realise that there is a cost associated with each of these choices. For example, in retaining a risk there is a cost associated with the contingency plans which are needed to manage it; and in transferring or sharing a risk, there is the premium charged by the party who takes the risk. Therefore, it is essential that everything be done to reduce or eliminate a risk before it is retained, transferred or shared. Not doing so will result in higher costs than necessary, assuming of course that the costs of reduction or elimination are less than the costs of transfer. If this is not the case, then simply transferring a risk to another party without any attempt at reduction or elimination might be the most efficient risk management strategy.

Although there are some risks that must be retained by law, the decision of whether to retain, transfer or share residual risks will depend largely upon the relative costs and benefits of each option and on an organisation's attitude towards risk. For example, a risk-seeking organisation would be more likely to retain residual risks than transfer them to another party. Nevertheless, many organisations fail to appreciate and assess the implications of retaining, transferring or sharing risks, and to help avoid this the following sections discuss the basic principles which should be followed in each case.

5.3.3.1 Risk retention

In risk retention, the potential losses attributable to a risk are directly absorbed by the organisation according to its risk appetite, which should be explicitly stated in its risk management policy. The benefits of retaining a risk are dependent upon the financial health and managerial capabilities of an organisation in being able to control it if it eventuated. The costs involved in retaining a risk are associated with the contingency plans that have to be put in place to reduce the possibility of it occurring in the first place and to minimise the consequences if it did occur. This of course assumes that an organisation is aware of the risks it is taking. In many instances, organisations unknowingly retain risks and fail to make the necessary allowances in their contingency plans. Unfortunately, all too often, the consequence if these risks then eventuate is conflict, as companies try retrospectively to off-load responsibility for the inevitable financial losses incurred.

In an ideal world, the costs of contingency planning can be recovered from customers. However, in reality, customer cost constraints, competition and general market conditions ensure that these costs often have to be absorbed. Furthermore, it appears that contingency allowances might be owned by customers whether they have paid for them or not. For example, a common contingency mechanism used by building contractors for potential time delays is to build "float" into a project's programme.

However, a recent court case in the UK has found that this float can prejudice a contractor's claim to time-related costs, even when the delay was caused by the employer (Ascon Contracting *v*. Alfred McAlpine Construction Isle of man 1997, ORB-361 and 1998, ORB-315). This means that when a contractor makes a contingency allowance in a programme, it provides protection for others as well as itself. Clearly, this is an important principle which must be considered when deciding whether to retain a risk or transfer it to another party.

5.3.3.2   Risk transfer

The main benefit of transferring a risk to another party is that the risk of the risk increasing is eliminated. However, in transferring risks there is also a cost, which is the premium charged by the party who takes the risk and the opportunity cost associated with the loss of benefit from potential opportunities.

Put simply, the two options in transferring a risk are: to transfer it to an *external company* that specialises in buying risks; or to transfer it to some *business partner* within a project. Both of these options are discussed below, in more detail.

TRANSFERRING RISKS TO EXTERNAL COMPANIES – INSURANCE

A useful, although unachievable, aspiration would be that all project risks would be capable of being distributed between members of a project. This would enable risk to be controlled in the optimum way within the skills, abilities and experiences of the project team. However, in reality it is inevitable that some risks will be beyond the control and financial capacity of the project team and will have to be handled by an insurance company. While insurance is necessary in most projects, too many organisations over rely upon it as a substitute for good risk management. In reality, insurable risks only represent a very small proportion of the total risks facing an organisation (maybe as low as 20 per cent). For those risks that can be controlled through effective management processes, insurance is the most high risk, inefficient and expensive option (Fenton-Jones 2003). Insurance should only be used as a last resort, for genuine residual risks which resist any other form of control. It certainly should not be the first option. With spiralling insurance premiums, particularly in the years following September 11 2001, the collapse of insurance giants around the world, the financial reporting debacles in the US and recent scandals in the insurance sector (*The Economist* 2004), those companies that have relied on insurance as a form of risk management are now counting the costs. With continued turmoil in the markets following these events, many companies now face a confronting choice of being driven out of business by the increasing costs of insurance or completely reviewing their risk management practices.

Although insurance should be a risk management strategy of last resort, it is undeniably a very important issue for the smooth running of any construction project and it is important to give it the attention it deserves. On many projects, insurances are arranged automatically with little thought and it is only when something goes wrong, when everyone begins scrutinising their insurance policies that deficiencies in cover start to emerge. To avoid such problems, we look at insurance in more detail in Section 4.3.4.

TRANSFERRING RISKS TO EXTERNAL COMPANIES – DIVESTMENT

All companies should evaluate the assets they own on a comparison of the relative risks and benefits they pose. In high risk/low return assets, the best treatment may be to divest – sell the asset to another business that may evaluate the asset in a low risk/high return manner. The decision in selling an asset should not be made before the potential risks to a purchaser have been mitigated otherwise this will be needlessly reflected in a reduction of the market value of the asset. For example, if a contractor owned a building on a contaminated site, the asset should not be sold until a site contamination investigation had been undertaken to define the scope and magnitude of the liability, followed by remediation work to remove the contamination. Of course, the reduction in value of not treating the site needs to be balanced with the up front costs of remediation before any decision is made.

TRANSFERRING RISKS TO BUSINESS PARTNERS

The main advantage of transferring a risk to a business partner rather than an external company is that it provides an incentive to manage it effectively, resulting in a better quality of service. To ensure that this decision is made wisely, there are a number of simple rules that should be followed. These principles apply to risk transfers between all business partners in a project, including employers, contactors, sub-contractors and suppliers. They are that risks should only be taken by those who:

- Have been made fully aware of the risks they are taking.
- Have the necessary capacity (expertise and authority) to avoid, minimise, monitor and control a risk.
- Have the necessary resources to cope with the risk eventuating.
- Have the necessary risk attitude (utility) to want to take the level of risk on.
- Parties accepting a risk should be able to charge an appropriate premium for taking it.

Not following these simple principles will result in confused responsibility for the vast array of project risks and create the illusion of risk transfer.

The ultimate result, when a problem arises, will be conflict, as parties argue that the responsibility lies with someone else. It must therefore be realised that everyone's fortunes in a project are ultimately linked and that one party can never fully transfer risks to another in a supply chain of which both are part.

Unfortunately, there is considerable evidence to suggest that risk transfer is often handled poorly between business partners in projects. For example, in the construction industry, Arndt and Maguire (1999) found that all too often the distribution of risk is influenced more by economics, commercial requirements, debt financier's requirements, bargaining power and company culture and policies than by the principles identified above. Furthermore, there is considerable research to indicate a poor appreciation of the importance of contract and procurement strategy in determining appropriate risk distributions between project partners. There is evidence to indicate that choosing the wrong approach can have serious implications for project success. For example, in the construction industry, *management-based procurement systems*, such as Construction Management, place the majority of risks with the employer as do *cost-reimbursement* contracts. These are not suitable for inexperienced, risk-averse, resource-constrained employers who do not have a reliable management infrastructure in place to monitor and control risks. For such an employer, a *Design and Build* procurement system using a *lump-sum* contract would be far more likely to result in a successful project by freezing the project price within contractually agreed constraints and minimising the management effort required to administer the project.

However, despite the existence of standard contracts to distribute project risks, contractual risk distribution is not simple. First, within contracts the main mechanism used to distribute risks are *express terms* written into the contract and *implied terms* automatically incorporated by common law. Therefore, being up-to-date with the frequent changes in common law decisions regarding a particular contract is essential to fully understand what risks are being imparted when it is used.

Second, to transfer a risk fully, it is necessary to consider all the possible consequences of a potential event and to make appropriate provision for them. For example, consider a situation where an express term in a contract is used to transfer the risk of adverse ground conditions to a contractor but ground investigation data is provided by the employer. Here, another express clause may be needed to avoid the possibility of part of the risk being transferred back to the employer (Uff 1995). It is clear that the way that risks are distributed under the law is complex and it is not surprising that contractual terms often do not clearly indicate which party is to be regarded as bearing a risk. This results in different perceptions of risk and misunderstandings about risk responsibilities, which are often exacerbated by the complexity of contracts in terms of their language and structure (Hartman *et al.* 1998).

A third problem which complicates risk distribution within contracts is that they have failed to keep pace with changes in roles, relationships and responsibilities within the construction industry (Trench 1991). Most construction contracts have their origins in the nineteenth century when the contractor carried out the majority of the buildings works. However, today the vast proportion of physical construction work is carried out by sub-contractors, meaning that risks that were once the responsibility of a single organisation are now the responsibility of many. This fragmentation of risk takers has produced a number of problems. First, contractors have tended to pass untenable risks to sub-contractors who do not have the resources, experience and expertise to manage them. Second, there have been problems of compatibility in the risk management procedures, interests and attitudes of the many sub-contractors that are involved in projects. Finally, as the number of actors on each project has increased, so have the boundary problems between them. The result, when problems arise, is the increasingly common excuse that everyone thought that everyone else was responsible for it.

5.3.3.3   Risk sharing

When a risk is transferred in its entirety to another party, there is an unfortunate tendency to forget about it. This is dangerous because in reality there are very few risks which are completely under the control of one party and all risk takers are interdependent to some degree. Therefore, wherever possible, risks should be shared between project members to develop the sense of collective responsibility which is needed to manage them effectively. Indeed, although it is not commonly appreciated, in reality this is always the case because the complexity of contract terms is such that risk is rarely placed fully or simply on one party (Uff 1995). Nevertheless, we consider below a range of mechanisms which can be used by decision-makers to share the risks associated with their decisions.

5.3.4   Mechanisms for sharing risks

Other than making explicit or implied provisions in contract clauses, there are other ways of sharing risks in business. These are discussed in more detail below.

5.3.4.1   Gain-sharing

Risk sharing should not be solely associated with the sharing of threats but also with the sharing of opportunities. To this end, we turn to the concept of "gain-sharing" which has become increasingly popular in many industries, although not in construction. Gain-sharing represents a family of pay-for-performance approaches that link group-wide financial rewards to employee and/or business partner created improvements in organisational performance. In this way, employees and/or business partners can share the risks of relative success or failure (Gomez-Mejia *et al.* 2000). While target contracts used in the construction industry facilitate the sharing of threats and opportunities in this way, gain-sharing differs in three main ways. First, gain-sharing links a portion of employee's or business partner's normal compensation to the achievement of specific goals in areas such as productivity, customer service, quality and any other goals included in gain formulas. Second, by design, any gains must be directly traceable to employee or business partner intervention and formulas only including those components which employees or business partners can affect directly. Third, rewards are shared with all members of a gain-sharing unit or group, regardless of individual contribution towards improvements. Fourth, rewards are provided which compensate employees or business partners for any reduced personal benefits associated with suggested improvements. Finally, the pursuit of gains in productivity involves a sequential process whereby suggestions for improvement are generated, assessed and implemented by employees or business partners with little managerial intervention.

Gain-sharing can only be used effectively when a number of conditions exist, namely, where employees or business partners have information about potential enhancements to productivity, where managers trust employees and where one can rely upon the cooperation of those employed in the production process. In highly unionised environments such as the Australian construction industry, some of these conditions might be difficult to achieve at employee level. Indeed, in many companies, without fundamental changes to traditional contractual and procurement practices, they may also be difficult to achieve with business partners.

5.3.4.2 Insurance

The risks transferred to insurance companies are sometimes called *actuarial risks* and are typically of very low likelihood and very high impact. Examples include events such as storms, fires, building collapses, fatalities, personal injuries etc. The business of an insurance company is to underwrite such risks for a premium linked to the probability and potential consequences of the event. This premium consists of a charge for covering the insured event for a specified sum (the policy limit) and a fee for profit, contingencies and overhead costs. Although such premiums can be substantial, they are invariably less than the potential loss which an organisation could face from an uninsured event and risk averse organisations will be happy to sacrifice a reduction in the NPV of their project for the improved predicability of potential losses.

Insurance premiums can be reduced by taking an insurance policy with specified *deductibles* which require the insured party to cover losses up to a certain amount, while the insurer covers all or part of the losses in excess of this, up to a policy limit. As well as reduced premiums, the added advantage of this system, which is commonly used in industries such as construction, is in providing greater motivation for the insured to manage the risk. However, this will involve a cost and, before accepting a deductible policy, a careful comparison must be made between the premium savings and the potential costs involved in managing the increased risks involved in accepting it.

The aim in securing any insurance policy is to gain the broadest cover at the most economic price. Whatever the type of policy undertaken, its cover must be for the full reinstatement value of the works insured, including fees. Indeed, in some cases it might need to exceed the contract sum if reinstatement costs involve the demolition of damaged property etc. Evidence indicates that these calculations are problematic for many risk managers in the construction industry. For example, Odeyinka (2000) found that on average, insurance cover in the construction industry falls short of actual losses by up to 40 per cent. Brown (1998) points to similar levels of under-insurance in other industries. He found that while the main kinds of losses are nearly always covered, it is coverage for the smaller and more obscure losses that tend to be missed. For example, many small companies are highly dependent upon steady cash-flows for survival but neglect to take out credit insurance in the case of customers or clients going bankrupt.

The most effective way of obtaining appropriate cover is to negotiate a custom-made policy with an insurance company with which an effective relationship has been developed over time. This is particularly important on projects where abnormal liabilities might exist. For example, in many Design and Build projects, standard insurance policies do not protect a contractor's liability for design, which can often exceed the legal obligation of architects and engineers providing design services to the project (Niemeyer 1998). Although the insurance industry is becoming more innovative and flexible in designing custom-made policies, this demands a proactive approach in facilitating the early involvement of insurance companies in the project planning and risk management process. Unfortunately, the proactive involvement of insurance specialists in construction projects is not common and the necessary data is often missing. Consequently, project insurance in the construction industry is normally arranged by a combination of standard *all-risk policies, multi-risk policies* and *special-risk policies*. All-risk policies are the most common. However, despite its title, an all-risk policy does not cover all project risks but a broad array of risks except those few which are excluded by specific insurance clauses. On the other hand, multi-risk policies specify a catalogue of risks such as fire damage, storm damage, death of workers etc. Unlike all-risk policies, only the specified risks are covered and unspecified risks are not. Finally, special-risk policies differ from all-risk and multi–risk policies in that only one specified risk may be covered.

While the range of coverage within these standard policy categories is complex and different for every project depending upon specific risks and insurance products on the market, the main types of insurance found on construction projects are as follows.

- Professional indemnity (PI): legal liability caused by negligence in carrying out professional duties for third parties.
- Construction all risks (CAR): insurance against damage to physical assets used during construction such as plant and materials, including any existing buildings or structures to be retained.
- Third party (public) liability: insurance against legal liability to third parties for physical injury or damage to property.
- Employer's liability: insurance which covers the employer against legal liability towards employees if they suffer physical injury or health problems while in their employment.
- Advance loss of profit: insurance against revenue shortfalls to all project participants from physical loss or damage under a CAR policy.
- Cap insurance: an additional insurance layer that covers the cost of an event which is outside the primary insurance layer – usually specified at a maximum cost per year.
- Environmental impairment liability insurance: covers the cost of remediation of environmental impairment from gradual events.
- Industrial and special risk insurance: covers first party repair to damage and third party property and environmental damage due to sudden events.

It would seem that the simplest approach is for the various parties to organise their own insurance. However, due to potential problems of overlapping insurances creating unnecessary expense and complexity, it is normal practice for the head contractor to arrange project insurance on a joint basis. More recently, the idea of "project insurance" is being advocated by the UK's Movement for Innovation (M4i) which was set up to implement the Egan Report. In this arrangement, the insurance company becomes a key member of the project team, issuing a single insurance policy covering the entire project and all the firms involved in it. This insurance continues for the life of a building, protecting the owner against any latent defects (Madine 2002). The advantages of this approach are: a reduction in insurance costs, the encouragement of collective responsibility for the management of risks and the involvement of an insurance company which monitors the risk management practices of participating firms. A further advantage of joint policies is that it removes the risk of *subrogation* claims against the insured parties by insurance companies. Subrogation is a general principle of law that allows the insurer, having paid a claim, to take-over any legal rights of the insured person against the third party who caused the original loss claimed for. However, there is no right of subrogation against people who are jointly insured under the same policy.

To conclude this section, it is important to point out that having taken out insurance for a specific risk, it is wrong to believe that it absolves the insured from the responsibility for managing it. First, if joint insurances are not taken, insurers who have to indemnify a party for loss are entitled in law to take action against the responsible party for negligent acts. Second, before payments for damages are made, insurance companies will want to know that everything was done to mitigate the risk and that there was no contribution through the negligence of the insured. Third, according to Odeyinka (2000) the average insurance pay-out typically only covers 60 per cent of the financial costs of an insured event. This is important because if insurance cover is inadequate or if insurers decide to dispute a claim, one is still liable to pay for any contractual responsibilities. Finally, insurance policies cannot compensate for the indirect costs of an event arising from issues such as damaged employee morale and tarnished customer relations. So there is little doubt that over-relying on insurance is dangerous. While insurance is an important part of any project, it is only a small part of the whole risk management process.

5.3.4.3 Securitised risk-transfer bonds

Transferring risks to private investors via bond markets is an alternative to insurance, which can spare risk mangers the trouble, expense and uncertainty of protracted disputes with insurance companies about definitions, triggers and coverage within policies. The key difference between a traditional insurance policy and a securitised risk-transfer bond is that the latter involves trading a potential loss on the capital markets by a bond issue. Every aspect of the bond transaction must be documented in a prospectus outlining the risk behind a proposed bond issue, trigger events which will release the bond payment (their severity, frequency and location) and how the loss will be treated (Friedman 2000).

The main advantage of securitised bonds over insurance is that payment does not depend upon proof of direct loss. However, securitised bonds are a one-off deal because once the bond is paid, the exposure remains, requiring another bond issue to cover its next occurrence. Unlike insurance, there is no on-going relationship with insurers who are expected to continue providing coverage after a loss. Furthermore, while disputes with insurers are less likely, there is no guarantee against legal disputes surrounding pay-outs on bonds, particularly when prospectuses are badly drafted.

5.3.4.4　Captive insurance

Another way of minimising problems with insurance is for large companies to form an internal *captive* insurance company to cover risks which are uninsurable by normal means. A captive is a subsidiary of a noninsurance firm that handles all or part of the parent's insurance needs – potential losses falling squarely on the shareholders' shoulders. In effect this self-insurance is a form of risk retention and has become more popular, especially with large contractors who are able to shoulder the burden of major potential costs. The advantages of captive insurance are:

- It reflects confidence in one's own performance, management capabilities and risk management systems.
- It provides greater motivation to manage risk – because it is a form of self-insurance.
- It overcomes the reluctance of insurance companies to provide cover for some types of risk.
- It avoids the fact that insurance premiums include an element of profit and fees of no benefit to the insured.
- It avoids excessive insurance premiums which make companies uncompetitive.
- It avoids excessive time-lags between premium payments and insurance pay-outs.
- It avoids effective performance, management capabilities and control systems not being reflected in reduced insurance premiums.
- It avoids the opportunity cost associated with the potential investment opportunities of unnecessarily high insurance premiums.

5.3.4.5　Mutual insurance

A variant of captive insurance is mutual insurance where a number of firms come together and agree to pay voluntary premiums into a joint account, which is then collectively, or independently administered to cover unforseen events included in the scheme. In the UK, a number of architectural practices faced with spiralling insurance premiums have formed such accounts which they then use to reinsure all or some of their risks at wholesale price (Button 2002). However, to be successful, there must be considerable trust between the parties and the scheme needs to be effectively managed.

5.3.4.6　Retention

Retention provisions are used in many construction contracts and sub-contracts around the world. Essentially, an employer withholds a percentage of each payment creating a fund for immediate use in the event of contractor insolvency or failure to remedy minor defects after the work is finished. In most instances, a pre-defined fund limit is set and a choice is provided between the principal being able to retain any interest or for the funds to be held in trust. Although retention is often talked about in terms of self-insurance for employers, strictly speaking it is not a form of insurance in that no third party is underwriting a risk for a fee.

The percentage retained is usually 3–5 per cent depending on contract value, although it can range from 1 to 15 per cent (Hughes *et al.* 2000). In most instances, half the retention is repaid on practical completion and the other half, when the holder is satisfied that contractual require- ments have been fully complied with. However, in reality, retentions have often proved less than effective. For example, in the case of serious defects, the amounts withheld rarely cover the costs of correction. Conversely, for simple snagging defects, the retention fund is too excessive and there is no incentive for a contractor to avoid such prob- lems and put them right. This is because on many occasions of good performance, contractors know that retention funds are sometimes never released or are withheld for years after the end of a project due to disputes or simple abuses of power by employers. Sub-contractors are particularly disadvantaged because even if retentions are released on time, some sub-contractors who complete their work early in a project might have to wait for years for the completion of the entire project before their final account is settled (Hughes *et al.* 2000). Sub- contractors are also disadvantaged in that the insolvency of a contractor makes it very difficult for them to obtain their retention, unless special arrangements have been made in their contract. Finally, independent trust funds are not always used and the retention monies are sometimes used to boost the holder's cash flows, putting it at risk. These abuses of retention cause severe hardships for the many highly geared, under- capitalised sub-contractors which constitute the majority of the construction industry and the effect is often financial hardship and even bankruptcy. Indeed, it has been estimated that retention can tie up as much as 20 per cent of a company's turnover and the total costs for the UK construction industry have been put at over 1 billion pounds per year (Cook 1999). As a result there is less money in the construction market for reinvestment in areas like safety, sustainability and quality.

5.3.4.7 Retention bonds

For the above reasons, the efficacy of cash retention is being questioned in favour of alternatives such as *retention bonds*, sometimes called *bank guarantees*. It is a system which is used widely in Australia where companies are underwritten for possible defects claims by a bank or insurance company (Davenport 2000). In essence, the guarantee is an agreement by the bank to pay monies up to the value of the guarantee, to the employer or head contractor on application to the bank by the employer. Bank guarantees/retention bonds may be conditional or unconditional (on-demand), most being the latter type. In an unconditional guarantee, the amount of the guarantee is payable on demand without the principal's need to establish default or to ascertain the losses suffered. However, in the conditional guarantee, the provider of the bond is only liable on the happening of a particular event and pay-outs are linked to being able to prove an event and the losses accruing from it. Unfortunately, the proving of this event and any losses can often be a cause of dispute between the principal, bond provider and bank.

Although retention bonds might be more costly than retentions because of the bank charges incurred in securing a bond, such a system is useful for increasing the industry's working capital and cash flows. It is also useful for screening the financial viability of a contractor or sub-contractor before they are employed because a bank's decision to issue a bond will be based upon the analysis of the contractor's normal overdraft limit, character, reliability, technical skills and management ability to undertake the specified work and resources available (personnel, machines and cash flow management). However, to work effectively it needs to be implemented uniformly across a whole project. For example, it is unreasonable for employers to allow their head contractors to provide retention bonds, if the head contractor is deducting retentions from their sub-contractors in the traditional manner. This is the main reason why its use in the UK is restricted to special initiatives by some sub-contracting associations and employers such as the Ministry of Defence. However, when it is used, there are some simple principles to follow to help the system work efficiently (Latham 1994, Klein 1997).

- Ensure that the overall value of the retention bond does not exceed 5 per cent of the contract value.
- Ensure that the bond reduces by 50 per cent at practical completion or even better at, periodic milestones throughout a project.
- Ensure that the value of the bond builds up in the same way as cash retention, otherwise the full value of the bond would be needlessly available on day one.
- Be vigilant if asked for both performance bonds and retention bonds because there can be costly overlaps of cover.
- Finally, although retention and bank guarantees provide some form of security for employers, there is often a shortfall in the coverage and other forms of protection/insurance will be required.

5.3.4.8 Surety bonds

Surety bonds are sometimes called *contract bonds* and are provided by a contractor or sub-contractor (principal) to an employer (obligee) as a guarantee that the terms of their contract will be fulfilled. Their coverage is thus far wider than retention bonds. In the US, surety bonds are mandatory on all public projects and commonly used on many private sector projects. In contrast, the UK and Australia have no mandatory requirement to use surety bonds although some employers do require them.

A surety bond normally specifies a limit to the liability of the surety, which varies between 50 and 100 per cent of the contract price. They are normally provided by an insurance company or bank (surety) that specialises in providing bonds. The advantage of banks is that they often provide a more speedy resolution of claims. A disadvantage is that surety tends to be offered as an extension of normal credit allowances, reducing a contractor's ability to borrow money. Ironically, this can also affect their performance on the project. While insurance companies are the most common source of surety, it is important to note that bonding differs from insurance because upon payment of a bond the surety has the right to sue the principal for the amount paid out. Furthermore, since the risk to a surety can be significant, individual sureties might not have the capacity to issue a bond on very large contracts and often limit their capitalisation, working in unison with several other co-sureties.

Four types of surety bond are commonly used on projects, namely *bid bonds, performance bonds, payment bonds* and *maintenance bonds*. The *bid bond*, which is sometimes called a *tender bond*, is issued to give assurances that the contractor will enter into a binding contract and will provide the required payment bonds, performance bonds and maintenance bonds if the contract is awarded. The surety agrees to pay the employer's damages if this does not happen. For example, the payment might be for the cost difference to the next highest bidder up to the face value of the bond which is normally about 5–20 per cent of the contract sum.

In contrast, *performance bonds* assure that a financially responsible party will stand behind the prime contractor (principal) if it does not perform in accordance with the contract. In the UK and Australia, performance bonds are usually limited to 5–10 per cent of the contract sum, although, with the emergence of PFI and PPP projects, performance bonds for 50 per cent, 80 per cent and even 100 per cent of the contract price have become common (Stebbings 2000). These bonds differ from normal performance bonds in that they include step-in rights which allow the insurer to step into the contract and assume the role of the original contractor.

Payment bonds give protection to an employer (obligee) if sub-contractors and suppliers are not paid by the prime contractor (principal). The sub-contractors are paid by the surety, although sub-subcontractors or suppliers to sub-contracts are not normally covered unless specifically written into the terms of a bond. The face value of a payment bond normally depends on the project value and are normally up to 50 per cent of the contract price.

Finally, *maintenance bonds* are used to provide funds to correct building faults discovered after construction is completed. Performance bonds, payment bonds and retention bonds terminate once construction has finished and become converted into maintenance bonds.

The main disadvantage of the surety system is the direct cost it adds to a project. On the other hand, with the assurance of being paid through payment bonds, tenders tend to be lower, which can reduce costs significantly. Indeed, the surety system reduces the risks of construction activity for many parties. For example, since a surety takes considerable risks, they usually conduct thorough investigations of a project including the background of principals and obligees, their liabilities and capabilities, the adequacy of working capital and credit, assets, reputations, personal integrity of key personnel, volume of work currently undertaken, experience in a particular type of work, how the project is planned, monitored and controlled, insurances taken, conditions of contract, solvency etc. For employers, the research undertaken by a surety is usually so thorough that there is often no need for a separate pre-qualification process by the employer. This can speed up the construction process. For contractors, risks are also reduced because the surety company will scrutinise contract terms and the qualifications of their sub-contractors. Consequently, this system can overcome many of the problems associated with contractual risk distribution discussed earlier. However, there is currently no established market for US style surety bonds in many countries which, compared to the bank guarantee system, requires far more involvement in a project by the surety and more specialist skills in risk assessment and in project appraisal.

5.3.4.9 Guarantees

Guarantees are often used to offset risks and can be classified into five main types, namely *supplier guarantees, seller guarantees, buyer guarantees, parent company guarantees* and *government guarantees.*

Supplier guarantees are useful on projects which depend upon highly specialist materials from specific suppliers. The success of such projects often depends upon suppliers guaranteeing to construct and operate special facilities to supply these special materials at a certain cost and by a certain time. Similarly, if a project depends upon a company or person parting with certain resources such as land or equipment then a *seller guarantee* might be necessary from them to control the price and time of sale. Conversely, *buyer guarantees* may be required from other parties such as tenants to ensure that they will purchase or lease a building upon its completion. However, by far the most common form of guarantee is the *parent company guarantee* which is used on many projects, particularly during times of recession, to ensure that the obligations of a subsidiary company will be underwritten by its holding or parent company in a financially stable group. While one would hope that good corporate governance would ensure that a parent company would not allow a subsidiary to collapse, leaving third parties out of pocket, it can happen. Parent company guarantees are designed to provide the same cover as a performance bond with the advantage of having no apparent cost for the employer or limit which may not cover the employer's cost of non-performance. However, nothing is free and the inevitable cost of parent company guarantees are merely hidden in the bidder's overheads and are not discernible in the tender price. Furthermore, a parent company guarantee might not provide as much protection against poor performance as it might at first seem, there being many general principles of law affecting a guarantor's liability. For example, a guarantor's liability might be discharged if there is a material change to the contract such as a simple variation. Also, if the guarantor is anything below the top holding company in a group's family tree, then supporting assets may be of little or only temporary significance and may be vulnerable to group restructuring. Furthermore, the *co-extensive* principle ensures that the guarantor is deemed to have the same liabilities as the contractor under the contract, which means that non-performance leading to termination of the contractor's employment also negates the guarantee. Nevertheless, it is probably fair to say that an employer gets better value for money out of a parent company guarantee than a performance bond.

Finally, we come to *government guarantees*. These have become common as governments around the world have embraced the principle of privatisation by opening up their infrastructure projects to the private sector through BOT, PPP and PFI type procurement systems. This normally involves some kind of concession contract which enables consortiums who build and finance a project to recover their costs with a profit by operating a facility over a specified period of time (the concession period). For example, the consortium which built Australia's largest ever privately funded infrastructure project, the $1.2 billion Melbourne City Link (MCL), was given a 34-year concession period to operate the private toll road. While this project was completed on time, such projects are highly risky. They normally have long payback periods and are often undertaken overseas in politically, socially and economically unstable environments by large custom-made joint venture companies. It is not uncommon for consortiums to find that they have to renegotiate concession periods to make projects viable. For this reason, consortiums often request some form of government guarantee backed by a multilateral institution against risks like changes in government policies, breaches of contract by local private and state-owned companies, low demand for the final facility etc. For example, the Colombian government agreed to reimburse the consortium which constructed the recent El Cortijo-El Vino Toll road in Colombia if traffic was less than 90 per cent of a specified level and went even further in the new runway at Bogota's El Dorado Airport, guaranteeing a minimum revenue.

Government guarantees are most common in countries which do not have strong government policies, a good and non-political regulatory framework and an independent judiciary. Without these, external investment in such projects is unlikely. However, it is unfortunate that such guarantees often undermine the benefits of projects to the local economy by reducing the incentive of private investors to manage them effectively and by exposing the host country's taxpayers to excessive risks. For example, in the 1970s, the Spanish government guaranteed 75 per cent of the loans on its new highway network and assumed the full exchange-rate risk, a decision that cost the Spanish Taxpayer US$2.7 billion.

5.3.4.10   Warranties and third party risks

With the development of PFI-, PPP- and BOT-type contracts, our perception of projects is changing dramatically to encompass its entire life cycle, from inception to refurbishment, to demolition and even beyond to reconstruction. The result has been some startling discoveries. For example, it has become more evident that the capital cost of constructing a typical office building can represent as little as 2 per cent of its total cost over 50 years (DHC 1980). This has huge implications for risk managers because it means that the largest proportion of risks associated with construction activity are taken by those who own, manage and use the building after it has been completed. However, until the 1990s, users of buildings and other distant third parties such as future tenants, future purchasers and financiers were insulated from the contractual framework of a project. Under the law of *Privity of Contract* (which means that a person can only enforce a contract if they are party to it), these third parties had to bear the risk of poor project performance themselves, the only recourse being a claim of duty of care through common law which was cumbersome, time-consuming and difficult. This led to the development of *collateral warranties* which brought third parties, who may find themselves liable for the costs of repair of a structure into direct contractual relationships with designers and, latterly, suppliers, sub-contractors, contractors and other consultants involved in the procurement of a building. Typical beneficiaries of collateral warranties are financiers, purchasers of a freehold or long leasehold or tenants who will occupy a building under a full repair covenant (RICS 2003). Collateral warranties give contractual rights to these people who would not normally be party to a contract. The term "collateral" means that its conditions should be closely aligned with your normal appointment conditions, and set no greater obligations than those set out within it. The original intention of collateral warranties was to allow third parties to sue for damages, but step-in rights which allow the warrantee greater latitude to continue administering a contract have emerged in recent years. For example, copyright clauses giving rights over designs are now more common and some warrantees may even require access to the staff of the warrantor to get critical information and to continue the contract. Throughout the 1990s, the range of warrantees became bewildering and it was not until 1995 that the first standard forms began emerging, although they have never been extensively used.

The need to provide a range of onerous, costly and sometimes dangerous collateral warranties with one's work introduced a whole new array of longer-term risks which had to be managed forcing people to think about the long-term consequences of their actions. While this was positive, a proliferation of warrantees within industries such as construction caused considerable confusion and, to overcome this problem, profound changes have recently been made to the law of Privity of Contract in the UK, for example, through the Contracts (Rights of Third parties) Act 1999. This act enables third parties such as tenants, purchasers and financers to stand in the shoes of a party to a contract and sue project participants, which in theory absolves the need for collateral warranties (Bingham 2000). Similar legislation has existed in New Zealand for some time. However, in the UK, third parties can only sue under a contract to which they are not a party, in two circumstances: First, when a contract expressly provides that an outsider can do so; and second, when a term of the contract "purports" to confer a benefit on a third party. So in circumstances where these require-ments are not satisfied or the conferred benefits are expressly limited, third parties will continue to rely upon collateral warranties. Indeed, anecdotal evidence indicates that the take-up by third parties, of the rights under the act has been very low. For example, Parisotti (2001) indicated that as many as 90 per cent of contracts used in the construction industry contain exclusion clauses to prevent the act being applicable. The main reason appears to be the added complexity, time and cost of including such rights. Furthermore, there is less direct protection than is afforded by collateral warrantees.

5.3.4.11 Financing risks

With the emerging popularity of PFI, PPP and BOT contracts, financial institutions are playing an increasingly prominent role in risk management. These contractual arrangements fundamentally change the traditional approach to the provision of infrastructure projects by the public sector, the underlying rationale being to combine the resources and risk management capabilities of the private and public sectors in the quest to provide more efficient public projects (Akintoye *et al.* 2001). Although there are many variants of these contracts, they are all based on the idea that the host government grants a licence to a private sector sponsor (normally a consortium), which will be responsible for financing, building, operating and maintaining an infrastructure project for a set period of time (the concession period). At the end of this period, the fully operational project is transferred back to the host government, usually at a nominal or zero cost (Walker and Smith 1995). Facilities normally considered for such arrangements include power, gas and water networks, sewerage facilities, air, sea and transport communication networks, and facilities for educational, health and correctional services (Arndt and Maguire 1999). The government normally requires that such projects are tendered on a competitive basis and specifies a minimum level of service to be measured against specific KPIs. Payments from the client then depend upon the attainment of this service level, an arrangement which transfers most of the risks to the private sector until the end of the concession period.

For the private sector, the very large size of these projects and the length of time over which risk exposure extends make risk management a critical issue. This is also a critical issue for the financiers of such schemes, whose involvement in constructing an innovative financial structure is critical to project success. Every project will be supported by some element of "project finance" where the financer considers the asset and revenues of a project in securing and servicing a loan. Strict non-recourse finance, where the financer has no claim on the non-project assets of the borrowers, is now rare and most financers insist on some risk being borne by the sponsors themselves. However, it is normally an expectation of project finance that the revenue flows from a project are enough to service any debt. Therefore, most financers need to be convinced that there is an efficient system for managing risks on a project and that risks and rewards are appropriately balanced. So far, evidence indicates that the lack of experience of PPP and PFI projects has often led to sub-optimal risk allocations and management strategies and the result has been a number of high-profile disasters (see case study of Sydney Airport rail link later in this book). As Arndt and Maguire (1999) point out, experience around the world has indicated that negotiations regarding risk allocations in PPP projects have often relied on a system of ambit claims, which leads to time-consuming and expensive negotiations where participants on both sides have not always understood their respective objectives and perceptions of risk. This has led to significant and unnecessary transaction costs, frustration among participants and diminished incentives for investment in new projects.

Finance risks are a particular issue in BOT, PPP and PFI projects and while banks initially financed most schemes, capital markets are now playing a greater role in them. Some of the tools developed by such markets to manage financial risks include derivatives such as *forwards* and *futures*, *options* and *swaps* (Akintoye *et al.* 2001).

Essentially, *derivatives* provide protection against adverse movements in prices and rates by fixing their future transactional value. This process is known as "hedging" and, in the case of *futures and forwards*, is an obligation to buy or sell an asset at a specified forward price at a known date. In effect, it is a risk sharing technique which creates a current cash flow in place of a future cash flow. An *option* provides the holder of a future derivative, a choice of whether to use it or disregard it and, like forwards and futures, can be traded on money markets. In contrast, *swaps* cannot be traded and involve the selling and purchasing of positive and negative cash flows associated with a scheme. It must be recognised that while hedges reduce potential losses, they can also reduce potential profits. For example, in a rising market for a service, a highly hedged service provider may not fully benefit from higher future prices and therefore may not perform as well as an unhedged provider.

5.4 Implementing, monitoring and reviewing risk and opportunity responses

Once a decision has been made about how to respond to a risk or opportunity, it will have to be implemented, monitored and, if needed, reviewed to ensure it achieves its stated objectives. These are the final stages of the risk management process and the purpose is to ensure that:

- Each person involved in the implementation of a response decision does as planned.
- Response decisions have the desired impact.
- Employees and external stakeholders are kept informed of progress towards implementation and resolution.
- Further risks or opportunities that may arise during implementation are detected and responded to.
- Lessons are learned for future risk and opportunity management.

5.4.1 Implementation

The implementation of any response decision will take time and may meet with resistance. Decision-makers will need to plan this process carefully and evaluate people's competencies in performing tasks, supervising them according to their capabilities and assessing the level of risk associated with their tasks.

To implement a response effectively, decision-makers will need to:

- Communicate the decision effectively to everyone who will be involved or affected – where necessary, in languages other than English.

- Create a clear timetable and implementation plan with specific milestones.
- Specify responsibilities for actions, giving responsibility to those best able to control the relevant risks and opportunities.
- Set clear and measurable objectives so that people know what is required of them and when implementation has been achieved successfully.
- Allocate appropriate resources to bring about required change.
- Put in place reporting/monitoring procedures to identify further risks and opportunities that may arise during implementation.
- Assist in the resolution of unforseen problems.
- Be prepared to adapt plans if needed.

All of these issues should be addressed in a "Risk Action Plan" which should act as a common reference point for everyone involved in the implementation process.

While a risk action plan is needed in responding proactively to risks and opportunities, in urgent situations, when reacting to risks or opportunities which have already arisen, a manager may first have to instigate emergency/ crisis management procedures. For this reason, every project should have an emergency response/crisis management plan which should be commensurate with the nature and hazards of the workplace, the size and location of the workplace and the number and mobility of persons at the place of work. An effective emergency plan should allocate and communicate responsibilities for the control of emergency situations. It should also identify a command centre, emergency contacts, requirements for the notification of authorities, detail training provisions, evacuation procedures, plant shut-down procedures, emergency equipment-testing procedures, medical and security provisions, public relations and media communications, investigation procedures and recovery plans etc. To ensure familiarity with these provisions, emergency procedures need to be clearly documented, communicated and rehearsed regularly. This process is discussed in more detail in Section 4.5.

5.4.2 Monitoring

Response decisions need to be monitored against the objectives described in a Risk Action Plan. This should occur in three ways:

- An initial review of the implementation process within a short period following the introduction of any controls.
- Using the reactive risk and opportunity identification techniques that form part of the normal risk identification process (see Section 2.6).
- Using "Risk and Opportunity Diaries" kept by each person named in the Risk Action Plan and submitted to regular risk review meetings (see Section 2.6.3). These are designed to ensure that specific actions have been taken to control risks and that they have had their desired affect.

5.4.3   Risk reviews

Response decisions should be subject to regular reviews:

- When monitoring procedures indicate that controls put in place by response decisions may not be effective or could be more effective.
- When monitoring procedures identify changes in the business environment or workplace that have the capacity to alter the type and level of risk or opportunity and the effectiveness of existing controls.

These unforseen events represent potential risks and opportunities and should initiate a new cycle of risk identification, risk analysis and risk control.

5.4.4   Learning

It is important to take time to reflect on the lessons learnt in making a response decision. Therefore, when a decision's objectives have been achieved, decision-makers should arrange a de-brief with stakeholders where this can be discussed. These deliberations should be recorded and become part of a project "post-mortem" which will record lessons learnt from every significant decision made over the life of a contract. This in turn should be stored in a company post-mortem which should record lessons learnt from every project a company is involved in. Of course, creating post-mortems is of no use if the lessons are not discussed and disseminated to improve project or company practices. Therefore, strategies should be developed to ensure that this process occurs.

5.5   Crisis management

In Chapter 1 we discussed how to judge an organisation's vulnerability to crises. The method we developed applied to projects and companies. Since then, we have constructed a proactive and preventative approach to risk management designed to prevent crises happening. Nevertheless, while prevention is better than cure, it is impossible for managers to create a risk-free environment and it is essential that managers have reactive strategies in place to deal with crises if and when they arise. This has become abundantly obvious in the context of the tumultuous events which have affected businesses in recent years.

Crises are a special type of extreme risk which manifests themselves in defining periods of acute difficulty which can threaten the existence of an organisation, its business units or its key products and even the lives of people. They tend to be low probability, high impact, unpredictable events which create the need for critical and rapid analytical decision-making skills, the results of which are likely to fall under extensive public, media and/or government scrutiny. They are also characterised by an extreme sense of urgency which hyper-extend an organisation's coping capabilities, producing stress and anxiety among organisational actors and stakeholders. If managed badly, crises can bring down companies and result in death, injury and destruction. Most worryingly, once they have occurred crisis appear to activate in-built defence mechanisms thus creating conditions which make their mismanagement more likely (Loosemore 2000).

While traditionally most managers have seen crisis management as a sign of failure, a necessary but unproductive and negative activity which can take an inordinate amount of time, there is an increasing realisation that crises are becoming an inevitable and often healthy part of organisational life, which should be planned for (Pascale 1991, Frazer and Hippel 1996, Furze and Gale 1996, Lerbinger 1997). This planning should be part of an integrated and thoroughly implemented risk management process. While crises can destroy unprepared organisations they can strengthen those that are well prepared. Nevertheless, in construction and facilities management, evidence suggests that many organisations exist in a low state of crisis preparedness, having an inadequate understanding of their risk exposure, of how to mitigate those risks and of the internal systems needed to cope with and learn and recover from their eventuality (Loosemore and Teo 2002). To improve this situation, the following sections discuss the essential elements of a well-conceived crisis management strategy. The same principles apply, whether we are considering the management of company or project organisations.

5.5.1 Crisis management planning

The importance of a well-conceived crisis management plan cannot be overstated as one of the defining characteristics of a crisis-prepared organisation. This has been illustrated many times. For example, one of the main contributors to Occidental Piper Alpha disaster was the almost complete lack of appropriate operating manuals on how to interrupt the potentially catastrophic sequence of events that led to the fire spreading (Bea 1994).

The main benefit of having a preconceived plan that can be automatically implemented when a crisis strikes is that it takes away some of the initial pressure and shock associated with it. This creates a valuable "breathing-space" within which people can calmly investigate the problem and agree on an appropriate response. During a crisis, every second counts and the first few hours are particularly critical. This is especially true if external constituencies are involved because initial impressions play a disproportionately large role in shaping their judgements of competence and blame. If initial impressions are bad then an organisation will be judged guilty until proven innocent and in many instances this can intensify a crisis and accelerate its escalation.

5.5.2 Disaster committees

Many organisations in high-risk industries have a permanent disaster committee that is responsible for championing the need for crisis management, identifying current preparedness and vulnerabilities, devising disaster plans, and coordinating people during a crisis (Kutner 1996). The membership of such committees is an important factor in determining their ability to do this. Ideally a disaster committee should consist of senior managers, managers from all functional departments and external professionals who have experience of crisis management, public relations, the law, and physical and mental health issues. Commitment from the top of an organisation is especially important if the activities of a disaster committee are to be taken seriously and if they are to have a chance of success.

5.5.3 Conducting crisis audits and creating crisis portfolios

A crisis audit assesses an organisation's crisis capabilities and identifies the inherent risk factors in its environment, internal activities, technology, infrastructure, and culture that need to be addressed to improve its crisis preparedness (Mitroff and Pearson 1993). A generic tool for doing this was developed in Chapter 1. However, it must be remembered that every organisation has different vulnerabilities, so the first stage in this process should be to develop a working definition of a crisis from its particular perspective. This will involve identifying and ranking, in terms of probability and consequence, the *types* of crises to which an organisation is vulnerable. This in turn will require managers to learn from past events and to look into the future to explore unusual combinations of events that may seem unlikely, but which could combine to produce a serious crisis. Ranking allows appropriate judgements to be made about the relative costs and benefits of constructing a plan for each possible crisis because limited resources ensure that planning for every possible crisis is not possible or indeed sensible.

5.5.4 Establishing monitoring systems and standard operating procedures

One aspect of the disaster committee's job is to establish monitoring systems to detect potential crises. This should involve the allocation of clear responsibilities for the monitoring and communication of all internal and external risks and opportunities. However, as we pointed out in Chapter 2, this is easier said than done in projects because workforces are highly transient, risks and opportunities continually change, and contractual relationships are often dynamic, complex and unclear. Nevertheless, every attempt must be made to ensure the widest possible coverage of potential risks and opportunities and to encourage collective responsibility for their monitoring and reporting.

The disaster committee should also develop standard operating procedures that define precisely *who* should be involved in a crisis response, *what* they should be doing, *when* they should be doing it and *how* they should be doing it. These procedures establish a pre-defined emergency communication network that needs to be followed during the early but critical phases of a crisis when people are disorientated by events. The intention is to "buy" the organisation some time to come to terms with events, to allow people to re-orientate themselves, and to ensure that appropriate resources are mobilized quickly and that they are commensurate with a crisis' scale. To do this, the procedures should be achievable, simple, flexible, and understandable by all internal and external stakeholders. For example, in many countries such as Australia, Singapore and the European Union, construction sites have many migrant workers and this may require the production of manuals in a range of different languages.

5.5.5 Creating a command centre

During a crisis, information is constantly being generated from a multitude of sources and it is critical that it is supplied "live" to the correct place, at the correct time and in an understandable format. In this sense, a key aspect of a disaster committee's job during a crisis is to identify a clear command centre that represents a single point of responsibility for decision-making and information management. A command centre is a critical coordination mechanism that helps facilitate a unified crisis management response since one of the greatest problems that can emerge during a crisis is the tendency for people to act independently. For example, in the case of a fire emergency, the command centre should have the sole responsibility to contact emergency services and to coordinate individual supervisors who are charged with clearing certain areas of the site. In contrast, in the case of an economic crisis such as the bankruptcy of a major sub-contractor, the command centre should be responsible for reorganizing work and re-employing another sub-contractor. In addition to being of practical importance during a crisis, command centres also play an important symbolic role. For example, Nicodemus (1997) discusses the fortunes of a company that faced a crisis and named their command centre "the war room", where they symbolically declared war on the problem.

5.5.6 Security

"Security" is another important issue for a disaster committee to consider. During a crisis, interference from unwanted elements can exacerbate the situation or, at the very least, interfere with its management. This involves identifying external constituencies who feel that they have a stake in the outcome of a crisis but who cannot contribute to its solution. While all stakeholders must be managed and kept informed, those involved in crisis management efforts should be insulated from these disruptive elements so they can develop a strong focus on the problem.

In some situations, it is also important that the site of a crisis is physically cut-off from potential disruptive elements, particularly when it continues to represent a danger to the public. In such situations, evacuation procedures may need implementing and it is essential that they are clearly communicated to everyone on a project and reinforced by regular training and mock-drills. For example, public address systems, sirens and horns can be used to notify people of an incident if they are placed at strategic locations so everyone can hear them. Whatever signals are used, they must be as simple and as unequivocal as possible. Also, responsibilities for using them must be clear, as should appropriate back-up if, as Murphy's Law dictates, key people are away on the day of an incident or if essential equipment malfunctions. A particularly important part of evacuation is the clear labelling of exit routes from all parts of a site. For example, people should know that mechanical hoists cannot be used in an evacuation and that all potentially dangerous machinery in the vicinity of escape routes must be switched off. Furthermore, since most project sites are a constantly changing physical environment, the positions of notices and their maintenance needs constant monitoring. It must be continually ensured that all evacuation routes follow the shortest possible route to checkpoints where role-calls can be taken in safety. They should also be wide enough to facilitate an orderly evacuation of the facility. For example, on an inner city construction site, this may be the street, and if this is the case the hazards to the public, to traffic and to site workers must also be assessed in consultation with public services such as the police.

5.5.7 Developing a culture of collective responsibility

The need to insulate a disaster response team from unwanted elements does not mean that it should be allowed to become introverted. Consideration also needs to be given to the reorganisation of non-crisis management activities so that the remainder of the organisation can function as normally as possible. Crises inevitably drain a considerable amount of energy from other functional areas within an organisation, demanding special efforts from the people who operate there. Clearly, without a considerable degree of goodwill and a sense of collective responsibility from unaffected parts of the organisation, the impact of a crisis could worsen and spread. Such goodwill cannot be expected if it did not exist before, and in this sense the crisis management process needs to be continuous and extensive in scope.

One way of developing a culture of collective responsibility is to communicate everyone's interdependency during a crisis. It also helps to clarify responsibilities for risks and ideally share them as much as possible. Most crises demand an injection of extra resources into a project and if the disaster committee does not identify in advance where this will come from, then a crisis will stimulate negotiations and potential conflicts that could delay a response.

Decisions concerning risk distribution are particularly relevant to economic crises, and earlier we argued that they have been a major cause of conflict within construction projects. This is particularly the case when the principles of risk transfer identified in Section 5.3.3 are not followed. These principles apply at all points along the contractual chain – to clients, consultants, head contractors, sub-contractors and suppliers. For example, if a contractor is employed under a high-risk contract by a client and has not been given the opportunity to price for those risks, then it is likely that it will attempt to transfer those risks along the chain by using back-to-back contracts with their sub-contractors. Indeed, sub-contractors may do the same and so on, until all project risks have been dissipated to the end of the contractual chain. Unfortunately, it is here that the most vulnerable, crisis-prone organisations exist. Inevitably, when problems that demand extra resources begin to occur, the end result of this risk-cascade is a backlash of conflict up the contractual chain as parties deny any responsibility for them.

5.5.8   Public relations

Public relations are an essential aspect of crisis management since most crises have implications beyond an organisation's boundaries. In essence, the three "publics" that need to be involved in a crisis are *employees* not directly affected by it, *external and quasi-external interest groups* and the *general public*. As Aspery (1993: 18) argues, "crisis communications built on well-established relationships with key audiences stand a better chance of protecting, even enhancing your reputation during difficult times. A company which decides to start communicating during a crisis will have little credibility."

5.5.9   The media

Throughout this book we have pointed to the increasing scrutiny of industry as a result of the ever-greater appreciation of its impact on the natural environment. This is particularly true of industries like construction which also have a direct affect on the built environment and therefore on most people's lives (Moodley and Preece 1996). This, coupled with growing public sympathies for the environmental movement, has resulted in increasing numbers of confrontations with the public, particularly on large infrastructure, mining and housing projects. For example, in the UK, protestors against road projects have tunnelled under proposed sites and organised highly effective publicity campaigns against those companies involved. Notably, in many of these increasingly common and public confrontations, the media has often portrayed construction companies in a heavy-handed and unsympathetic light and there is little doubt that the future viability of many projects and the image of the industry will have been seriously tarnished by this coverage.

The negative media coverage which construction has received is not surprising given that many construction companies attached little importance to the building of sound relationships with the media. Many companies see public relations as a non-value-adding activity and perceive journalists as dangerous, untrustworthy and irresponsible (Moodley and Preece 1996). The perceived rejection of the public and media has been particularly strong during crises when construction companies have appeared to consciously hide from the public, considering them to be an unnecessary distraction from recovery efforts. Yet this is precisely the time when it is most dangerous to ignore the media since in the after-math of a crisis, the public has a tendency to embark on a process of ritual damnation. This is particularly true of high-profile, publicly financed projects in which people may feel a greater right to recrimina-tion as a result of having paid their taxes to finance it. Public relations is clearly an area where the construction industry must improve is perform-ance. Most construction companies have long underestimated the power of the media in shaping public opinion and it is remarkable, given the environmental impact of the industry's activities, that many of the world's largest construction companies still do not have any in-house public relations expertise.

One way of ensuring open communication with the public during a crisis is to establish a 24-hour-a-day press office, which has the responsibility of pro-viding factual and up-to-date information to the media and to company employees. If managed well, such an office can turn media inquiries into opportunities rather than problems by initiating, rather than reacting to, press, radio and TV coverage. Experience of crises in other industries shows that public relations are best handled by one trained person who is named as an official spokesperson and who has skills in dealing with the media. TV interviews with untrained staff who appear uncaring, flustered and unsure of the facts are damaging to the public's perception of competence, whereas a trained person with experience of such events can portray a positive image.

5.5.10 Training

Many investigations into disasters such as the Piper Alpha fire have cited a lack of training, drills and exercises as a crucial contributor to the escalation of a crisis. Without training, people respond to crises in different ways and often irrationally, introducing further risks into the situation. It is unlikely that every person will be a good crisis manager and it is important that companies put in place training systems, supported by regular emergency drills and practices, to increase the predictability and consistency of the crisis response. This will ensure that when a crisis occurs, everyone knows what their roles and functions are, what the limits of their responsibility and authority are and how they fit into the overall response strategy. The first stage in this process is to decide if, when and where training is necessary. The second stage is to clearly document the standards of behaviour that are required from the various staff who are likely to be affected by a crisis – before it occurs. This must be done at all organisational levels and is particularly important for the crisis management team who must act as a cohesive coordinating unit during a crisis. The third stage is to decide on what type of training is most appropriate to different people. The fourth stage is to carry it out and the final stage is to monitor and evaluate its effectiveness.

An effective training strategy clearly delineates between different types of training and what they need to achieve. An effective training strategy is likely to include a combination of competency-based training (to teach a task or procedure), skills-based training (to teach a skill or behaviour), education (to teach someone to think) and exercises/drills/simulations (to practice or test a skill or procedure). The location and timing of the training is also important. For example, training can be delivered internally or externally, on-site or off-site, individually or team-based, face-to-face or distance (electronic), formal or informal, continuous or periodic, mandatory or voluntary, specialist or general etc. The challenge in developing a training programme is to combine these different options in a way that responds to the specific needs of every organisational function at every organisational level. The ultimate goal is to bring about permanent changes in people's behaviour and to this end it helps if all training is:

- Seen as worthwhile – relevant to people's needs and value adding
- Clear in its objectives
- Un-intimidating
- Enjoyable
- Realistically achievable in terms of end results desired
- Supported by peers, colleagues and superiors
- Convenient – cognisant of existing work and personal commitments
- Supported after the event by monitoring and feedback systems about implementation and performance – trainees must not be abandoned.

5.5.11 Post-crisis management

After a crisis, a disaster committee should organize follow-up meetings so lessons can be learned and fed into subsequent crisis management efforts.

Everyone affected by a crisis must be involved in this process. In addition to managing the learning process, the disaster committee should also turn its attention to the recovery. This can be a lengthy and sensitive process that is likely to be influenced by how well a crisis was managed. For example, it may involve delicate challenges such as conducting investigations into causes, mending damaged relationships, reorganizing the project programme, settling ongoing disputes and re-assessing project requirements. At the same time, attention must be given to the long-term consequences of a crisis such as rectifying damage to the environment, or dealing with government or legal investigations. Clearly, the less effectively a crisis is managed, the more arduous is the recovery process.

5.6   Recovery – Business continuity management (BCM)

Business continuity management is concerned with how an organisation plans proactively to re-establish key business processes in the aftermath of a crisis and to ensure survival in the longer term. This focus on business resilience and sustainability is in contrast to *crisis management* which is concerned with how an organisation copes with a crisis while it is happening. It is also different to *disaster recovery management* which is concerned with how an organisation survives in the short term, and to *contingency management* which is concerned with developing alternative back-up strategies in the event of existing system failure.

Clearly, BCM is a critically important aspect of any comprehensive risk management strategy and has a central role in minimising the longer-term social and economic impact of a crisis. However, according to International Data Corporation's most recent business survey of Australian businesses, while 83 per cent of companies with over 100 employees have a BCM plan in place, only half had tested it. In the public sector, only 32 per cent had tested their BCM plans. There is no reason to believe that businesses are any different elsewhere, although the New York Stock Exchange has recently proposed a rule which will require all listed companies to demonstrate that they have in place a business continuity plan, that they have tested it regularly and that they have a senior manager as its guardian. Similarly, in 2003 the Australian Stock Exchange introduced new ASX Corporate Governance Guidelines requiring similar reporting provisions, and in March 2004 the Australian banking and insurance sectors introduced new standards to ensure that their customers have effective and well-tested risk management and business continuity plans. The finance and IT sectors appear to be the leaders in BCM because of their heavy reliance on computerised data systems, and there is much that firms in other industries can learn from their practices. For example, facilities management companies could learn from companies like Hewlett-Packard which operate 44 recovery centres worldwide, housed in anonymous buildings, in undisclosed locations and away from city centres. They require customers to rehearse for crises on a regular basis so that they can be up and running as quickly as possible in the event of one. IBM is also interesting since they provide several recovery sites for customers to use in the event of a crisis so that they can continue their operations. IBM also strives to prepare and test for such crises before the need arises. Finally, The Reserve Bank of Australia requires its exchanges and clearinghouses to have back-up arrangements for critical communications and computer systems and key personnel. Indeed, most big banks now have their data warehouses located at least 20 kilometres outside metropolitan areas, but are often reluctant to reveal their exact locations and testing patterns for obvious reasons (Timson 2003).

The Australian National Audit Office has found that the most effective BCM programmes are implemented from the top down, being driven by senior managers who champion the cause. The first stage in developing a BCM programme is to develop a clear plan for development and implementation with key objectives and milestones. The next step is to ensure that managers understand their business – what its business is, what its goals are, what its culture is (values), who its key stakeholders are, what methods are used to achieve its goals, what its business plan is, what financial, human and technical resources are available, what the main risks and opportunities are etc. Having done this, the next step is to undertake a business impact analysis, which involves asking questions which revolve around "outage". These might include posing scenarios such as: what impact would a computer system crash have on the business? What if a key piece of machinery broke down? What if a key person left the organisation? Would we be able to continue in our business and survive? Such questions are designed to identify the maximum acceptable outage – the amount of time an organisation can last in the event of an outage, before having to instigate a BCM response. For example, a high technology organisation might only be able to last a few hours in the event of a computer crash while a construction company may be able to last a few days. Having identified maximum outages for different systems, technologies, people etc. the next step is to develop a treatment plan to mitigate potential outage losses. For example, in the event of a computer system crash, this may involve identifying and arranging an alternative source of computing support. Having developed treatment plans for each system outage, the penultimate step is to document them in a BCM Plan and to implement them, by making arrangements with internal and external stakeholders involved in a response. For example, in the event of a computer system crash, it may be necessary to make arrangements with external contractors and suppliers to temporarily take charge of certain key business functions. Having documented the BCM, the final step is to regularly audit, test, refine and maintain it, frequently identifying and managing potential internal and external barriers to its effective implementation in the event of a crisis.

5.7   Conclusion

In this chapter, we have discussed the various strategies that can be employed in effectively responding to a potential risk or opportunity. The guiding principles are simple and revolve around: *openness* and *participation* so that everyone is fully aware of the risks and opportunities they are exposed to; *equity* so that no party is forced to take a risk or opportunity they are uncomfortable with; *collaboration* so that parties accepting a risk or opportunity are able to charge an appropriate premium for taking it; and *expertise* in ensuring that those taking risks and opportunities have the necessary knowledge, systems and capabilities to manage them effectively. The potential benefits of adhering to these basic principles are enormous as are the potential dangers of not. If they are followed throughout the entire supply chain from the employer all the way down to the supplier, risks and opportunities will be distributed naturally from person to person in the most efficient manner. However, if they are not followed or if there is one weak link in the chain then there is a danger that some risks and opportunities will go unnoticed, unassessed and unmanaged. Then, the seeds of ineffective and inefficient responses will be sown. We end this chapter with a case study which illustrates the practical importance of following the principles discussed above. It explores the process of risk distribution which was employed on the Sydney Airport rail link BOOT project in Australia.

5.8   The Sydney Airport rail link

This case considers the process and pattern of risk distribution on the Sydney Airport rail link in Australia. We will also discuss the implications for the success of the project.

5.8.1   Introduction

The Sydney Airport rail link is a new 10 km underground railway comprising two tracks and four new stations, which provide rail services between Sydney Airport and the Sydney central business district. It is jointly owned and operated by the public and the private sectors. An illustration of the project is provided in Figure 5.1.

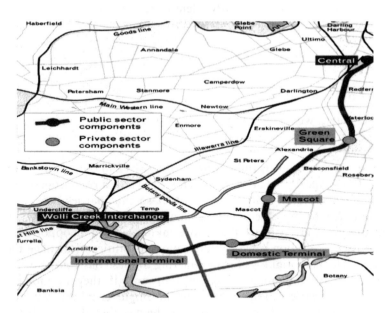

Figure 5.1 The Sydney Airport rail link.

5.8.2 Project summary

- Name of project: Sydney Airport rail link
- Name of private promoter: Joint venture of Transfield ALC Pty Ltd and French group Bouygues SA
- Type of delivery: BOOT
- Project cost: $920 million
- NSW Government funding: $704 million
- Shareholder equity: $30 million of $224 million private funding
- Financier: National Australia Bank
- Concession period: 30 years
- Construction start: June 1995
- Completion date: May 2000
- Actual construction time: 4 years and 11 months

5.8.3 Project benefits

As well as providing a rail link between Sydney's airport, the central business district and the wider rail network, it was anticipated that the Sydney Airport rail link would also:

- Reduce 25 per cent of road traffic between the city and the airport.
- Act as a catalyst for urban consolidation in an area that included three universities and recreational facilities such as beaches, parks and golf courses.
- Increase rail capacity between Erskineville and Sydenham, thereby deferring the Sydenham six-track extension and immediately releasing $60 million in reserves.
- Provide additional essential capacity for the East Hills-Campbelltown Corridor between Turrella and the central business district, thus facilitating transport links with the rapidly developing south-western suburbs and allowing the long-term deferral of track augmentation between Sydenham and Erskineville.
- Create at least 3000 jobs during construction, which will boost the economy.
- Support the objectives of government transport policies and the wishes of public transport activists and community green groups.
- Provide public transport infrastructure which would contribute to the stimulation of urban containment and renewal in the central industrial area, particularly Alexandria and Mascot.

5.8.4 Project history and organisation

Although the idea of a Sydney Airport link was proposed as far back as 1915, its genesis as a real project lies in an unsolicited BOOT bid by a consortium comprising CRI Ltd, Qantas and Westpac in 1990. The State Rail Authority (SRA) subsequently invited open tenders and received four bids. Of these, two were ultimately shortlisted, namely CRI/Quantas/Westpac and Transfield/Bouygues. In 1991, more detailed proposals were submitted to SRA and, after further discussions, a special purpose company named the "Airport Link Company" (ALC) was created to continue the development of the project. This was established by Transfield and Bouygues in an initial 50/50 share arrangement supported by the National Australia Bank with a debt-equity ratio of 86 per cent. Eventually, 100 per cent of shares would be acquired by Transfield.

In 1993, the ALC submitted a tender to design and construct the government-funded works (underground track and tunnel) for $473.64 million and to invest up to $30 million to finance, build and operate the private sector components of the project which comprised four stations. However, in 1994 political pressure caused the state government to terminate the project and re-seek competitive tenders. This caused a dispute between the private sector and the public sector partners which was resolved by an independent consultant who concluded that there had been no lack of probity and that the bid was deemed to be good value for money and contain an equitable sharing of risks. In response, SRA went forward with the project and embarked upon a final stage of non-competitive negotiations. Final contracts were signed in February 1995 between SRA and ALC with airport site access being provided by the Federal Airports Corporation. Construction began in June 1995 and was finished in May 2000, with the concession period for the operation and maintenance of the four stations, tunnels and tracks ending in 2030. The contract price was initially valued over $650 million, $484 million for tracks and tunnels (funded by SRA) and $128 million for the stations (funded by the private sector). However, due to major changes in scope, the project finally cost over $900 million, $704 million of this being paid by the public. The parties involved in this project are illustrated in Figure 5.2.

5.8.5 Financing

The final agreed financing structure involved the SRA funding its own tunnels, tracks, catenary, signalling, communications systems from its capital works budget (at a design and construction cost of $542 million). The state government also provided $133 million from consolidated revenue, despite a refusal by the Commonwealth Government to assist in funding the project by supplementing the SRA's Capital Works budget. Under a Stations Design and Construct Contract, ALC agreed to finance, build, own, operate

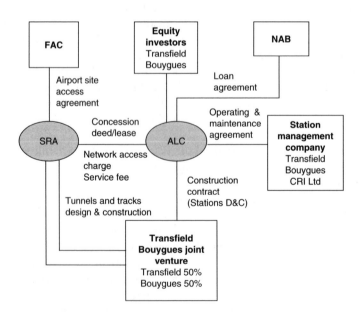

Figure 5.2 Parties to the agreement.

and transfer the four stations, and from 2000 to 2030 would lease the strata containing these stations from SRA's control. ALC would also design and construct the public sector component of the project (tunnels, track, signalling systems, power and communications), under a turnkey type "Tunnels and Tracks" Contract which was an agreement for ALC to operate, staff and maintain tunnels, tracks and associated infrastructure during the 30-year concession period. After this period, the stations will revert to the public sector. The operator was a specially created Station Management Company owned by Transfield, Bouygues and CRI. ALC would pay SRA a nominal lease payment for the station land and SRA would charge a network access fee which is payable after debt is recovered (scheduled to occur in 2012) and a train service fee (linked to patronage) which is payable from opening in 2000.

ALC's operating income is linked to revenue from retail activities at the stations and the fare structure which is in two parts:

- The train fare – which is the normal SRA fare as regulated by the State Government Pricing Tribunal, with children's and other concession type fares still applying.
- A station access fee – a tariff of $6 for the airport stations and $1 for the two other privately operated stations.

A special tax concession was necessary to improve the financial feasibility of the project, in which tax is only applicable after a certain amount of debt servicing and operating costs has been recovered. Debt servicing is anticipated to be complete in 2012.

5.8.6 The allocation of risks

Table 5.1 summarises the distribution of risks between public and private sector on this project.

5.8.7 The allocation of benefits

Table 5.2 summarises the distribution of potential benefits between the public and the private sector on this project.

5.8.8 Conclusion

The Sydney Airport link contract price was initially valued over $650 million, $484 million for tracks and tunnels (funded by SRA) and $128 million for the stations (funded by the private sector). However, due to major changes in scope, the project finally cost over $900 million and the cost to the public was $704 million. Furthermore, the ALC defaulted on a $200 million loan

Table 5.1 Distribution of risks between public and private sector

Stage	Risk bearer state rail authority	Airport link company
Pre-design	SRA took all approval risks. Necessary approvals were made a condition precedent to the contract. If not forthcoming from the Environmental Authority, the Australian Tax Office and local government planning requirements then ALC or SRA could have terminated or renegotiated the agreement. The airport link proposal was passed through five local government areas in addition to the airport, which is located on Commonwealth territory. Given the complexity of the approval process under the existing state environmental legislation, the Minister of Planning streamlined the decision-making process and formulated a new State Environmental Planning Policy.	None
Design	SRA carries the risks associated with delays or costs associated with dealing with the Federal Airports Corporation.	ALC provides full design for tracks, tunnels and station infrastructure for a lump-sum price. Design of stations are to be fit for intended use and must ensure efficient operation once completed. Since most of the tracks are tunnelled underground, the design risk was substantial.
Construction	SRA purchased land along the track route and bears risk of the site not being accessible as expected. SRA can use the government's powers of compulsory purchase if necessary. Should the site be delivered after the agreed time, the concession term can be extended or compensation can be paid. This issue is particularly important for the two stations at the airport which are sited on Commonwealth land administered by the Federal Airports Commission. The SRA has entered into a separate agreement with Federal Airports Commission to provide the land required for the stations and tunnels by certain dates.	Construction of stations, tracks, tunnels and associated infrastructure on time and within lump-sum price (including fixed inflation allowance) and to agreed level of quality.

	SRA bears risk of general industrial disputes aimed at government policy but not project-related disputes.	
	The risk associated with force majeure is excluded from the private sector.	
	SRA responsible for the provision of airline pedestrian links.	
Operation	SRA must provide train services in accordance with the train service standard.	ALC responsible for station operation and maintenance costs associated with tracks, tunnels, stations and associated infrastructure are borne by ALC. This is done within fixed station operating and maintenance costs agreed between ALC and SRA which will be reviewed on an "open book" basis every five years.
	SLA is responsible for operating trains, selling tickets and meeting agreed service standards.	ALC is reliant on the SRA operating trains on time and to a frequency as spelt out in the agreements.
Revenue/ financial	SRA does have some exposure to production cost risks because lower patronage results in a lower train service fee and network access fee to the ALC.	The cost of operating the stations is primarily an ALC risk since the amount which can be charged to use the stations is fixed.
		ALC has to pay a network access fee and a service fee to SRA, which are based on net revenues generated by ALC determined on a five-yearly basis. The greater the operating cost of the stations, the lower the network fee will be. To some extent this offsets the risk to ALC; however, since fees are only reset every five years, ALC bears the full production costs risk within this period.
		ALC carries the risk on changes to taxation, other than changes in sales tax and customs duty.
		ALC carries the risk of equity subscriptions and subordinated loans.

Table 5.1 (Continued)

Stage	Risk bearer state rail authority	Airport link company
	SRA carries risk of changes in requirements and changes in the law or government policies which directly or indirectly affect usage of airport. Compensation is provided. SRA's returns depend on the level of patronage due to the network access fees being linked to patronage levels. SRA also earns direct revenue from ticket sales to users of the NSR.	ALC carries ongoing market/ revenue risk over the 30-year concession period, since the level of revenue is directly dependent on level of patronage using the train line. Extensive modelling during the development phase in 1993 predicted patronage of 46,000 trips per day. Patronage was expected to increase to 65,000 per day in 2013 due to population growth and development of the south Sydney area for business and residential developments.
	Since revenues received by SRA in the form of the network access fee are contingent on the debt being paid off, SRA has some exposure to interest rate risk.	The risk of fluctuating interest rate has been allocated to ALC. ALC has hedged this risk by interest rate swaps to fix the interest rates for its loans.
		While predominantly funded in local currency, some of the major items used in the construction (such as the tunnel boring machine) are imported. Exchange rate risks have been allocated to the SRA where items are purchased under the Tunnels and Tracks contract.

barely six months after the line opened with patronage of only 12,000 passengers per day compared to the original estimate of 46,000 per day.

The main problem appears to have been that the $10 premium rail fare was above the affordable price and not well marketed. There was no apparent benefit and incentive to attract people to use the rail, which was in competition with alternative conventional transport mode such as bus and private vehicle. For example, the time between trains is between 7½ minutes and 10 minutes in peak period, and the journey from Sydney's Central Station to the international terminal takes 10 minutes. However, Sydney's airport is only 15 minutes from the city using the newly built Eastern Distributor and the taxi fare is around $20. The economics of the train therefore disappear if a patron shares a taxi with another person and there is also the added advantage of greater flexibility and convenience for the traveller.

If ALC's loan defaults to the National Australia Bank continue, the government may be forced to purchase the four privately built stations despite the government's preference for it to be privately financed. To avoid

Table 5.2 Distribution of potential benefits between public and private sector

Stage	Benefit bearer state rail authority	Airport link company
Pre-design	SRA benefits from initial private sector proposal.	ALC benefited from initial proposal lodged with SRA from other private sector consortium.
Design	SRA benefits from expertise of private sector and incentives to provide an innovative and cost-effective (capital and life cycle) design.	ALC can design the facilities to maximise their operating revenues.
Construction	All construction is performed by the private sector resulting in higher efficiency in project delivery and costs.	ALC can ensure correct construction time, cost, functionality and quality of facility to maximise their operating revenues during concession period.
Operation	All operation and maintenance of stations and tracks are undertaken by ALC.	ALC has a 30 year concession period agreement where is can learn to operate the facilities efficiently.
Revenue/ financial	Apart from land purchases and funding the design and construction of the new lines, SRA's capital outlay is small.	Through efficient design, construction and operations, ALC can maximise its income flow from the project.

this scenario, there may be a case for government intervention to boost patronage by encouraging the use of rail. The challenge is to modify people's behaviour which will demand mutual co-operation between all public and private stakeholders such as State Rail, Sydney Airports Corporation and the airline companies. Other options include concession fares to groups and multi-ticketing through travel agencies (combined airline and train ticket at special discounts).

The above problems have led to significant public criticism and questions surrounding the long-term viability of the Sydney Airport link project and the risks of BOOT projects in general. It would seem that the main dilemma facing governments on BOOT projects is the balance of risks between the public and the private sectors. The natural tendency is to shift as much risks as possible to the private sector. On the other hand, if the private promoter bears the majority of risks, the project may not be commercially viable because of the risk premium involved. Other important issues relate to: accurate market forecasts, supported by measures to change people's consumption behaviour; ensuring efficiency, transparency, equity and certainty in government tendering processes and contracts; ensuring clarity and stability in government objectives; ensuring the project sponsor's ability to raise the necessary funds and take the necessary risks; and reducing the complexity of contractual arrangements.

Chapter 6

Developing and implementing a successful risk and opportunity management system

6.1 Introduction

Only now, having discussed each stage of the risk and opportunity management process, are we in a position to discuss how to develop and implement a successful risk and opportunity management system. In creating such a system, organisations are faced with an array of approaches lying on a continuum from *informal* to *formal*, the appropriate choice depending upon the size and complexity of an organisation and of the projects it undertakes.

6.2 The informal approach

The informal approach is characterised by a relative absence of documented risk and opportunity management policies and processes and is suited to small organisations undertaking relatively low-risk, simple, repetitive projects in stable environments. Typically, procedures are unstructured and subjective, the output being no more than a limited set of contingency allowances added to project programmes and cost estimates. It is an approach which is adopted by many companies, even large ones, yet evidence suggests that it is inefficient, particularly in complex projects where the consequence is reactive management to unforseen problems resulting in possible crises, losses and delays.

6.3 The formal approach

The formal approach consists of a set of explicit procedures for organising the risk and opportunity management process. These are designed to become a routine and habitual aspect of project management, providing integrated guidelines for all organisational levels, enabling uniformity of approach and greater objectivity in decision-making. Ideally, the result is a flexible system which guides people through the stages of risk and opportunity management by motivating and prompting them to think about risks and by providing tools for identifying, analysing and responding to them.

The creation of a formal risk management system involves three main steps.

1   Creating and communicating a risk management policy
2   Creating and implementing a risk management system
3   Building a risk management ethic into corporate culture.

These stages are discussed in more detail below.

6.3.1   Creating and communicating a risk and opportunity management policy

A risk and opportunity management policy is a statement of intent which should communicate an organisation's attitude, rationale and philosophy towards risk and opportunity management. More specifically, it should perform the following functions:

- Make a promise to customers about achieving a certain standard of risk management and a commitment to deliver it.
- Communicate an organisation's priorities/objectives (critical success factors against which risk and opportunity management activities will be judged).
- Communicate the benefits of risk and opportunity management to everyone (at a personal level and by linking risk and opportunity management activities to corporate objectives and strategic plans).
- Communicate the need for risk and opportunity management and its importance.
- Communicate specific attributes of an organisation in terms of special risk and opportunity exposures.
- Be internally consistent with other policy statements.
- Be externally consistent with and acceptable to other organisations in a company's business environment and supply chains (e.g. Clients, sub-contractors, suppliers and regulatory bodies).
- Establish the goals and objectives of the risk and opportunity management function within an organisation.
- Clarify a general approach to risk and opportunity management and reasons for and expectations of risk and opportunity management activities (formal/informal and qualitative/quantitative etc.).
- Communicate the range of issues to which the risk and opportunity management policy applies.
- Define the duties, authority and responsibilities of the risk and opportunity manager.
- Define an organisational structure and management system to provide a motivational and operating framework for risk and opportunity management activities (communication channels, risk management

team/department, management information systems, training, education to support those responsible for managing risks and opportunities etc.).

- Define the relationships of the risk and opportunity management team/ department with other departments.
- Communicate a clear policy but be flexible enough to prevent people becoming slaves to a bureaucratic system.
- Be holistic in approach and not concern itself with prescribing detailed strategies, mechanisms and tools for identifying, analysing and responding to risks and opportunities at operational level.
- Be capable of revision and updating in response to changes in company objectives and its business environment.

Clearly, the creation of a risk and opportunity management policy is a serious undertaking that demands careful management over a realistic timescale. It should not be taken lightly or rushed. The significance of the above issues requires the policy statement to be formulated at senior executive level. There will be many major issues to be discussed, choices to be made and decisions to be communicated which may not have been formally and previously debated within a company.

Having created a policy, its effective communication is critical to ensure successful implementation. This can be facilitated by developing an action plan which involves:

- Establishing a team of risk and opportunity management champions, involving senior managers, to be responsible for internal and external communications about the policy.
- Initiating a consultation process with those who will eventually implement the policy to ensure ownership and commitment to its principles. This should include any stakeholder (external or internal) with which there is risk and opportunity interdependency, such as financiers, major clients, consultants, major sub-contractors, suppliers etc.
- Creating an education programme to raise awareness about the benefits of managing risks and opportunities.
- Creating a workable and practical risk management system that is suited to the risks and opportunities an organisation faces, its priorities and its human resource capabilities.

6.3.2 Creating and implementing a risk and opportunity management system

Many organizations, in their enthusiasm to embrace risk management, have made the mistake of developing unduly complex and intimidating systems that, while appearing rigorous, have had a negative rather than positive impact. These companies have forgotten that developing any management system should be a journey, not a destination, which should evolve in response to the nature of its activities and to the capabilities of its workforce, rather than the ambitious aspirations of its managers.

The steps in developing a risk and opportunity management system involve creating a:

- Risk and opportunity management team
- Risk and opportunity management manual
- Risk and opportunity management plan for each project
- Risk and opportunity management training and reward system
- Risk and opportunity management information system
- Risk and opportunity management performance measurement system.

Before progressing, the importance of consultation must be emphasised again. No organisation can isolate itself from the risks of other organisations in its supply chain. This requires that these business partners also go through the same steps and develop compatible risk and opportunity management systems. Only then will an organisation have developed a truly comprehensive system that provides maximum protection against its risks and maximum potential to capture its opportunities. Indeed, the law increasingly requires risk management systems to be developed in this consultative manner, especially in relation to health and safety risks (WorkCover 2001).

It is also important to point out that any risk and opportunity management system must be regularly reviewed and adjusted to maintain alignment with changes in an organisation's risk and opportunity profile, business environment, workforce capabilities, policies, priorities and objectives. To this end, an organisation should review its risk and opportunity management policies and systems on an annual basis and be prepared to respond on an *ad hoc* basis to major changes in any of the above factors.

6.3.2.1 The risk and opportunity management team

The creation of a risk and opportunity management team is an important resource decision which helps to communicate an organisation's commitment to its risk and opportunity management policy. An effective team should include a range of people with the necessary attributes and expertise to champion, drive, develop, monitor and continually improve the risk management process. In large organisations, this may involve the creation of a separate risk and opportunity management department or section. However, in smaller organisations, it might simply involve allocating these responsibilities to existing staff. The appropriate structure depends on the range and magnitude of risks and opportunities facing an organisation and the resources available to invest in risk and opportunity management activities. Nevertheless, it is probably true to say that the more power and resources given to this team, the more seriously the risk and opportunity management process will be taken.

The role of the risk and opportunity management team is to perform an integrating function between people in different functional departments in an organisation, assisting them to make decisions more effectively. The ultimate aim should be to nurture a culture where risk and opportunity management become an instinctive and automatic way of thinking for all employees. This will involve documenting, defining and communicating the roles and inter-relationships of everyone involved in the risk and opportunity management process.

A risk and opportunity management team should be led by a risk and opportunity manager who is responsible for championing risk and opportunity management, monitoring risk and opportunity management practices and reviewing the risk and opportunity management system to ensure it is responsive to company priorities and changes in the business environment. To be able to do this, the risk and opportunity manager must occupy a rel-atively senior position with involvement in strategic policy decisions. While a risk and opportunity manager might be responsible for the design and implementation of a risk and opportunity management system, it is a fatal yet common mistake to make them responsible for all risk and opportunity management activities. Systems centred around one individual can never work effectively and, ultimately, the success of any system must be the respon-sibility of everyone who has to make decisions. The risk and opportunity manager's job must be to champion the risk and opportunity management process and to ensure that the necessary infrastructure and support exists to advise, train, mentor and coach staff to make their decisions more effectively. In larger organisations the complex array of risks and opportunities norm-ally requires the support of an equally complex array of specialists. These specialists may include legal experts who can provide specialist advice on contractual and insurance provisions, computer experts who can use highly sophisticated risk simulation software and who can create and manage effective databases, expert facilitators who can help teams identify their risks and opportunities more effectively etc.

6.3.2.2   The risk and opportunity management manual

The risk and opportunity management manual is created by the risk and opportunity manager in consultation with key stakeholders and senior and operational managers. It should provide consistency in risk and opportunity management practices throughout an organisation and be a source of reference material for all decision-makers. More specifically, the risk and opportun-ity management manual should provide practical instructions and specific guidance for use in decision-making processes. This involves documenting the detailed techniques of risk and opportunity identification, analysis and response that an organisation considers best suited to its business and employees. The manual should also set clear standards, lay down measures of performance and describe audit and review procedures. Furthermore, it should list the names, positions and responsibilities of all persons who will have specific responsibilities for identifying, assessing and responding to risks and it should specify detailed rules/arrangements for ensuring compliance with various laws and organisational policies. Finally, the manual should also contain reference material such as the major provisions of insurance contracts, worker's compensation funds and employee benefit plans etc.

6.3.2.3   The risk and opportunity management plan

The risk and opportunity management plan is a project-specific document which acts as a record of data, decisions and deliberations made during each stage of the risk and opportunity management process, over the entire life cycle of a project. This information is recorded in a *risk and opportunity register* which lists every decision, its identified risks and opportunities, their likelihood and potential consequences in terms of project objectives, who owns the risk and the response strategy in each case (including identifying responsibilities for actions needed). As well as being a record of the risk and opportunity management process, the risk and opportunity register acts as a critical hazard control document for a project, recording, prompting and guiding managerial actions to deal with risks and opportunities. In this way, it provides a valuable audit trail and basis for status reports throughout a project and an important learning mechanism between projects. Learning is at the core of risk and opportunity management and risk and opportunity registers from one project can be very useful in providing lessons about effectively foreseeing, assessing and controlling risks and opportunities in others. Risk and opportunity management should not be seen as a linear process with a distinct beginning and an end but as a continuous cycle of improvement in working practices.

The risk management plan should also include *risk and opportunity diaries* completed by every person named in the risk management register as being responsible for implementing risk and opportunity management actions. The purpose of this document is to ensure that the specific actions to control risks and opportunities have been implemented and have had their desired effect. This is important because it is virtually certain that unforeseen problems will arise during implementation to prevent this from happening. For example, if the control of a risk or opportunity rests upon changing working practices then this could be met with industrial unrest and safety problems associated with unfamiliar work arrangements. In other words, risk and opportunity management creates its own risks and having a record of them is important to ensure that they are also responded to.

6.3.2.4   Training and reward systems

Training is the foundation of effective risk and opportunity management because any system, however sophisticated, is only as good as the people who are using it. Training not only teaches people the techniques of risk management but also plays an important motivational role by making people aware of the personal benefits of managing their risks and opportunities effectively.

Training can be provided at a number of levels depending upon the sophistication of the risk management techniques advocated by the risk management manual. For example, if training in sophisticated quantitative methods of analysis is justified then it is common for specialist software houses to provide the training needed to support their software. However, most managers do not need such complex training and much can be achieved with generalised awareness training that teaches simple qualitative techniques of identification and analysis. This can be done in-house through external consultants running short custom-made courses or by the use of internal mentoring and coaching systems. Alternatively, it can be done by putting staff through formal education programmes in colleges and universities. Whatever combination of training is chosen, the ultimate aim should be to achieve a permanent change in people's behaviour so that risk and opportunity management becomes imbedded in every decision involving significant risks and opportunities.

However, while it is easy to change someone's behaviour, to do so permanently is much more difficult. Therefore, it is important to ensure that the training is relevant and practical. In particular, it is essential not to be too ambitious in the early phases of developing a new risk and opportunity management system. Over time, as the knowledge, capability and confidence of an organisation's workforce improve, so can the sophistication of the training and the outputs expected. Finally, to bring about permanent behavioural change, it is important to support any training with rewards and incentives for people who successfully implement what they have learnt. This will require the creation of a performance monitoring and appraisal system which is "behaviour-based" to evaluate and reward those who have changed the way they work in a meaningful, beneficial and enduring way. In such a system, the emphasis should always be on positive reinforcement through rewards rather than negative reinforcement through punishment. While punishment might reduce poor behaviour, only rewards will encourage outstanding behaviour.

6.3.2.5   Information systems

Effective risk and opportunity management is completely dependent upon quality information which is accurate, up-to-date, easy to use, understandable and accessible. Therefore, the design, management and maintenance of high quality record keeping systems, databases and communication systems to supply information to the right place at the right time in a useable format is very important.

Once again, the nature of any information system depends largely on the nature of the risks and opportunities an organisation faces, its objectives, its priorities, the sophistication of its workforce and its policies relating to risk and opportunity management practices. There is little point is supplying complex statistical data relating to probability distributions if managers are not able to use it.

The data needed by decision-makers should enable them to effectively *identify, analyse* and *respond to* risks and opportunities. This should be both qualitative and quantitative, relating to the nature of potential opportunities and threats, their frequency, imminence, severity and consequences. Lessons from previous decisions are also important to know how effective various responses were in mitigating risks and maximising opportunities. The maintenance of accurate insurance premium and loss figures is important too, as are details of policies, contract law developments and information relating to the outcomes of past risk and opportunity response decisions.

Whatever data is provided to decision-makers, it is essential that it is of the highest quality and collected from reliable and independent sources. For example, much can be collected from market research, interviews with experienced staff, published statistical sources and stakeholders such as clients, suppliers, sub-contractors, research publications, pressure groups and statutory bodies. Good record keeping of past events is also critical and should be recorded as it becomes available to those involved in a project through risk registers and diaries (rather than retrospectively). Consequently, it is important that people understand and comply with their information collection responsibilities relating to risks and opportunities. Consequently, it is also important that people understand and fulfill their data collection responsibilities for managing risks and opportunities. Any records of past risks or opportunities must record the chain of events which precipitated them so that their causes can be tackled before potential problems become crises and opportunities get taken by competitors. By combining all of this information, a company can build up, over time, an excellent understanding of its unique risk and opportunity portfolio/profile and what response strategies tend to work best in different contexts.

Finally, there is little point having excellent information if it is not disseminated effectively. To assist in this process, it is useful to allocate responsibility for database management to a specialist in data collection and analysis. This person should have responsibility for collecting, analysing and disseminating information to the right people at right place at the right time and in a usable format. They will also need to ensure that any data is accurate, high quality, usable and up-to-date. Furthermore, they will need to ensure that any information system is responsive to individual needs which will necessitate consulting decision-makers about what information they need, when they need it, how often they need it and in what form they need it. Once again, methods of dissemination depend upon resources available and the sophistication of an organisation's workforce. Methods can vary from manual distribution to interactive company intranets. The advantage of intranets is that they can also carry instructions for different risk management techniques, examples of good practice and discussion forums where managers can highlight and debate their current problems and solutions.

6.3.2.6   Performance measurement system

Businesses are subject to constant change and it is essential that risk management is seen as a continuous, dynamic and responsive process. At an individual level, this requires that any decision to control a risk or an opportunity must be constantly monitored and reviewed. The creation of a performance measurement system is a critical part of this process because, ultimately, the success of a risk or opportunity management strategy must be measured by its impact upon the attainment of organisational objectives.

6.3.3   Building a risk and opportunity management ethic into corporate culture

Culture refers to the shared values and beliefs that members of an organisation have about what is important and about the way they should act, behave and relate to each other in doing their jobs. A culture permeates and influences everything an organisation stands for, does and communicates internally and externally. In terms of risk and opportunity management, it means encouraging people to see themselves, first and foremost, as risk and opportunity managers in a specific functional role, rather than vice versa.

Getting the culture right is extremely important and numerous corporate disasters have illustrated that if an organisation's culture does not support its strategy then the attainment of corporate objectives is unlikely. A vivid example was the sinking of the Herald of Free Enterprise where 188 people lost their lives. The official investigation revealed that "from bottom to top, the body corporate was infected with the disease of sloppiness...revealing a staggering complacency...individually and collectively they lacked a sense of responsibility" (Boyd 1990). The consequences for the board of directors was enormous with a verdict of unlawful killing being delivered after a fifteen-month police investigation leading to manslaughter charges against seven individuals and P&O European Ferries being charged with corporate manslaughter. This demonstrates that the management of corporate culture is critical if risk management strategy is to be successfully implemented. No risk management system, however sophisticated, can compensate for a weak risk management ethic in an organisation's culture. Ultimately, the challenge of risk management is to manage the fit between culture and strategy so that the organisation's goals can be achieved.

6.4   Conclusion – The risks of risk management

While there are many benefits to be gained from developing and implementing an effective risk and opportunity management system, there are also some risks which organisations should be aware of. These have the potential to more than undermine any benefits gained. Therefore, strategies need to be put in place to deal with them. For example, having invested so much time and energy in developing a new system, staff can become complacent and treat the system itself as a substitute for good risk management. The temptation is to think that the system itself will provide adequate protection from the unknown and that there is no need to do anything. However, good risk management is more an attitude of mind than simply having a system in place. The effectiveness of any system depends as much on people changing their behaviour as it does upon a well-designed document.

Another potential risk is that some people, in their enthusiasm to adopt the system may believe that they are invincible and take more risks than they did before – risks that perhaps they cannot manage. So people's behaviour needs to be monitored so that they work within their capabilities.

Another risk is that an effective risk and opportunity management system will tend to highlight more problems than before. Although this is a sign of the system's success, there is an accompanying danger that people become disillusioned with the process and see it as negative and confrontational. People may even start covering up problems, especially if they are to blame for them. To prevent this, managers need to instil a no-blame culture and encourage people to look for opportunities as well as risks. In particular, some early successes and measurable wins in identifying previously unknown risks and opportunities will go a long way towards providing momentum and an appreciation of the benefits of the system for users.

Another problem is the danger of paralysis-by-analysis, where a preoccupation with analysing decisions in great detail results in over-caution and hesitancy in taking risks or a slowing down of decision-making to the extent that opportunities get missed. This would be ironic since the system is designed to enhance opportunities. In this way, increased knowledge is in itself a significant risk. The way to avoid this problem is to ensure that the system is flexible and that people can use their judgement when using it. People need to know that the system is a tool not a straightjacket and that users and stakeholders in a decision should arrive at a consensus about how to use it. People should be encouraged to experiment with the system because it is only by doing this, that its strengths and weaknesses will be discovered and improvements be made. It is also important to communicate that the system is not meant to stifle risk taking. Indeed, an effective risk management system should enable people to take more risks – but they should be calculated risks rather than "leaps in the dark".

Finally, a common risk associated with any standard system of any kind is that people stop talking and assume that everyone else understands it in exactly the same way as they do. There is also the danger that people over-rely on the system and assume that it will do their job for them. Although training can overcome this problem to some extent, it is important that people understand that despite being standardised and applicable across a company and a supply chain, it is unlikely that everyone will understand it in the same way. Furthermore, every project is different and needs a different approach. Teams should therefore be encouraged to discuss their under-standing of the system before using it and to adapt it, if needs be, within certain limits so that its consistency and coherence is not destroyed. This chapter ends with a case study which illustrates the practical application of the issues discussed above. It looks at the process by which Multiplex Facilities Management (MFM) successfully developed and implemented a new system to effectively manage the risks and opportunities on its customers' facilities around the world. This major case study provides a fitting conclusion to this book since the system also integrates all the principles, ideas and techniques we have discussed in the preceding chapters.

6.5   Multiplex Facilities Management's new Risk and Opportunity Management System (ROMS)

The Multiplex Group is a listed international property investment, develop-ment and facilities management company, based in Australia, with an annual turnover of over AU\$2 billion. It is a major force in construction in Australia, New Zealand, Middle East and more recently the UK.

Multiplex Facilities Management (MFM) is a subsidiary of the Multiplex Group and was established in 1998 to offer developers, building owners and investors an integrated range of facilities and property management services. MFM extends the services offered by the Multiplex Group to the entire building life cycle – from inception, through design and construction to the ongoing use and management of the built facility. MFM's business is based on the increasing recognition by many property owners, that the land, buildings and support infrastructure owned by an organisation is a vital support resource that makes an important contribution to its customers' core business objectives. MFM's core business is to ensure that it effectively manages its customer's facility-related risks, thus ensuring that their real estate assets support and contribute to its core business objectives in an optimal manner.

6.5.1   The start of MFM's journey

We have said that risk and opportunity management is a journey, not a destination, and this case study is about an ongoing process that started in 2001. At that time, MFM had many good reasons to review and improve the effectiveness of its risk management procedures. For example, MFM existed in a relatively high-risk environment and were managing some of Australia's most prominent landmark buildings. The company had also grown rapidly since its formation in 1998 and the Directors were aware that they needed to keep updating their existing management systems to keep pace with this growth. In addition, MFM existed in an increasingly legislative and competitive business environment where the penalties for legislative non-compliance and contractual non-performance were becoming more severe. This was particularly so in occupational health and safety and environment management where a reputation for non-compliance could very easily destroy good customer relations and public relations, leading to a potential loss of business, increased insurance premiums and increased finance costs. In parallel with this trend towards increasing legislation, the world of corporate responsibility was also emerging fast, as the public became more sensitive to the impact of the facilities management industry's practices on their lives and more empowered to do something about it. This trend towards greater moral, ethical and social responsibility in business has continued and it is now estimated that over 35 per cent of all decisions made on the UK stock exchange are based on a company's non-financial performance. Similarly, the US Dow Jones Index has an in built sustainability index to guide investors and in Australia, the recently published ASX Corporate Governance Guidelines (the system by which Australian listed companies should be managed and directed) promote accountability and control systems to manage the social risks involved in achieving corporate objectives. While such guidelines only apply to listed companies at the moment, it is likely that they will be increasingly used and adapted by financiers in assessing the reputational risks associated with non-listed companies in the future.

In addition to the above, MFM were seeing the character of its customer base changing, its clients becoming more informed, more sophisticated, more demanding and more risk-averse. In particular, those large and potentially lucrative clients who themselves had sophisticated risk management systems were expecting the same standards of their business partners and were insisting on risk management audits as part of the prequalification process. MFM's future involvement in PPP and PFI projects presented a particular challenge with significantly greater levels of new risk being accepted over extremely long periods of time. Being an organisation that considered itself customer focussed, MFM inevitably found itself facing the prospect of taking new risks which were challenging to manage and control.

One of the major problems facing MFM, peculiar to the emerging profession of facility management, was the need to manage risks which had been passed "down the line" by contractors and designers, who did not understand the facility management business and over whom there had traditionally been little control. The Directors of MFM realised that, eventually, if risks and opportunities were going to be effectively and fully managed, they would also have to change practices along the entire supply chain that fed their business. This meant that any new risk and opportunity management system would have to be suitable for use in all stages of the building procurement process from inception, through to feasibility, design, planning, construction, facilities management and redevelopment and that extensive consultation from the outset with people involved in each stage would be necessary.

In summary, MFM saw risk and opportunity management as its core business function, a crucial and distinctive capability which would provide confidence to move forward into an increasingly competitive business environment with the ability to accept risks and opportunities that its counterparts did not have the confidence to manage. MFM also recognised that risk and opportunity management should be an integral part of good management practice and that the benefits could be substantial in terms of improved efficiency, better performance, increased competitiveness and higher profitability. Indeed, research at the time indicated that these opportunities were huge. For example, a survey of the UK's top 75 development companies in the construction industry indicated that 57 per cent regularly declined tenders on the grounds that projects were too risky (Smee 2002). Finally, MFM also knew that companies that were prepared to take a close and critical look at their risk management practices would be seen more favourably by insurers and financiers, potential and existing clients (particularly sophisticated clients), shareholders, pressure groups and the general public.

6.5.2   Auditing MFM's existing approach to risk and opportunity management

In response to the changes in its business environment, the Directors of MFM instigated an audit to establish the maturity of risk management practices within MFM. The review was based on the Project Management Institute's Risk Management Maturity Level audit tool, adapted and expanded for the construction industry (as explained in Section 1.6 and illustrated in Appendix A). This involved an analysis of MFM's business under the following headings:

- Awareness – How aware are MFM's employees and business partners about the importance of risk and opportunity management? How well is risk and opportunity management communicated throughout MFM and along its supply chain? How aware are employees, business partners and stakeholders of the importance attached to risk and opportunity management by MFM?

- Skills – What levels of skill do MFM's employees and business partners have in risk and opportunity management? What is the level of experience among staff and business partners of risk and opportunity management processes?

- Culture – To what extent is risk and opportunity management integrated into MFM's business culture and supply chain? Is it seen as a core activity? Do people *think* risks and opportunities instinctively? Do they cooperate to manage risks and opportunities? Is there strong leadership and championing of risk and opportunity management?

- Image – What is the external image of MFM in terms of risk and opportunity management? What are customers' perceptions of MFM's risk and opportunity management processes?

- Processes – To what extent do consistent and systematic risk and opportunity management processes exist in MFM and along its supply chain? Are contractual, procurement and management practices sympathetic to the principles of effective risk and opportunity management?

- Confidence – How confident are MFM employees and those in the supply chain about risk and opportunity management?

- Resources – To what extent are dedicated resources allocated to risk and opportunity management processes (e.g. training, support, risk management units/teams etc.)?

- Application – To what extent are risk and opportunity management processes and techniques applied to decision-making on a day-to-day basis? Is there application throughout MFM, from strategic to operational levels?

Answers to these questions and others allowed MFM to rate itself (on a scale of 1 to 4) in each of the above categories. This was then plotted onto a spider diagram (see Figure 6.1) to give a visual representation of MFM's strengths and weaknesses in risk and opportunity management. Figure 6.1 illustrates that MFM was operating between Level 1 and 2 in most categories. It shows that while most MFM employees practised risk management on a day-to-day basis, there was a need to invest in a new umbrella system, which could establish systematic and consistent procedures across the entire company and throughout its entire supply chain. The new system would need to instil a sense of collective responsibility for the proactive management of risks and opportunities throughout the entire company, encouraging an open flow of information between management and operating staff. This would help to release the entrepreneurial spirit of employees by encouraging people to show initiative. It would also allow MFM to

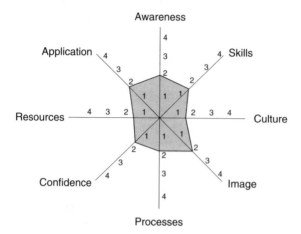

Figure 6.1 MFM's risk and opportunity maturity levels.

identify its major sources of risk and opportunity, to establish the best way to handle them and to assign clear responsibilities for managing and tracking them. The short-term challenge was to get the company operating at Level 3 in all categories and in the medium to long term, to achieve Level 4.

6.5.3 Clarifying objectives

MFM decided it wanted a formal but flexible system that would not burden people in complex procedures or stifle the gut feeling, which had been a traditional hallmark of Multiplex's success. In particular, it was important to MFM that the system reflected, communicated and reinforced its business culture. This meant that the system had to:

- Be customer-focussed and flexible – adaptable to the individual needs and circumstances of each MFM customer and add real value to their business objectives.
- Maximise client satisfaction – ensure MFM's clients' core business needs are met and ideally exceeded, through the efficient procurement and management of their projects and facilities.
- Demonstrate exceptional service – provide evidence of MFM's dedication to innovative work practices and exceptional service performance through system documentation.
- Ensure continual improvement – facilitate continual improvement in MFM's performance beyond expectations through vigilant and expedient response to all potential risks and opportunities. No one was to be immune, from the very top to the very bottom of the organisation.

- Be innovative and forward thinking – incorporating the latest thinking in risk and opportunity management and exceeding international standards (AS/NZS 4360: 1999 *Risk Management*).
- Realistic – bring about realistically achievable improvements in risk management practice within resource constraints and existing knowledge levels. A five-year time horizon was set to totally implement the system, bring about cultural change and develop an internal capability equal in sophistication to any company in the world in any industry.

They were also clear about what they *did not* want. In particular, they were wary of adopting an actuarial approach to risk management that was highly complex, quantitative and reliant on computer technology. They preferred a simple system which everyone would use to a complex system which would "sit on the shelf" and never be used. Furthermore, they did not want a system which burdened people with paperwork, rules and procedures. There was a determination not to create a paper chase and the Directors were acutely aware of the dangers of "paralysis by analysis" and of the need to be decisive and risk being wrong, rather than agonising at length and being right too late.

The ultimate objective of the system was to take the risk out of taking risks and to make risk and opportunity management an integral part of MFM's business culture, fundamentally changing people's behaviour so that risk and opportunity management became an instinctive and habitual way of working for every employee. In other words, MFM wanted their employees to "think" risks and opportunities. The Directors of MFM felt that if this could be achieved, then they would better harness the inherent capabilities of employees, a large proportion of whom were professional qualified.

6.5.4   A risk and opportunity management policy

Risk and opportunity management policy of MFM embodies the philosophy, values and objectives described in the previous section. A summary of this policy is presented in Figure 6.2 and it hangs on the wall of the head office reception area, visible to all employees, business partners and customers.

6.5.5   Establishing a unique approach to risk and opportunity management

An important step for any company embarking on the development of a new system is to identify the unique approach, which will distinguish it from its competitors. In order to establish and define a distinctive approach to risk and opportunity management, MFM realised that it had to understand and address current deficiencies in risk management in the construction and facilities management industries and in other high-risk industries.

MULTIPLEX FACILITIES MANAGEMENT

RISK AND OPPORTUNITY MANAGEMENT POLICY

MFM will work to mitigate risks and maximise opportunities to its customers, staff, business partners and society as a whole. The objective is to ensure that MFM's customers' core business needs are met and, ideally exceeded, through the efficient procurement and management of their facilities.

The philosophy underpinning MFM's policy is one of prevention rather than cure and, of collective responsibility for the efficient management of risks and opportunities. MFM will endeavour to provide a transparent, non-adversarial and open business environment in which this can be achieved and will empower its employees with the necessary skills, confidence and support to manage risks and opportunities effectively.

MFM's procedures for managing risks and opportunities are set out in its "Risk and Opportunity Management System" (ROMS). This provides a common framework for risk and opportunity management across the company with consistent terminology, tools and techniques. It is designed to guide people through the risk management process, yet be flexible enough to enable them to make individual choices to deal with risks according to their own specific circumstances.

This policy will be updated annually to ensure it is consistent with MFM's customers' objectives. The ROMS will also be updated annually to ensure that they comply with this policy statement.

Signed: _____ (Director)

Date: _____

Figure 6.2 Risk and opportunity management policy statement.

Therefore, a detailed review of systems and practices in the construction industry and in other industries such as mining, chemical processing, nuclear power, oil, gas, defence and steel was instigated. This review indicated that the main weaknesses of current approaches to risk management were as follows:

6.5.5.1   Negative rather than positive

Most approaches to risk management, despite the rhetoric of many companies' systems documentation, are not driven or inspired by the profit and value enhancing opportunities which risk management can offer (the upside of risk) but by the fear of the ever greater penalties for doing something wrong (the downside of risk). We have already discussed the reasons behind this negative mentality and there is no need to reiterate them here.

To address this common problem MFM decided that the crucially important difference between risks and opportunities would be a central and distinctive feature of its approach to risk management, which would distinguish it from its competitors. It was for this reason that the new system was entitled the Risk and Opportunity Management System (ROMS). This reflected a determination not to pay "lip service" to the concept of opportunism but to provide practical guidance and tools to actively encourage it amongst decision-makers. For example, the ROMS requires decision-makers to identify, analyse and maximise the opportunities associated with their decisions before thinking about risks and provides a vocabulary (excluding words like catastrophic) to enable them to do so.

6.5.5.2   Unsystematic and inconsistent

While most managers practice risk management on a day-to-day basis, it is often practised in a unsystematic and inconsistent manner. This means that standards vary considerably between and within companies and that risks go unnoticed and unmanaged. Like every management activity, risk management is best practised when it is underpinned by sound systematic and consistent processes. Without such processes, companies are too vulnerable to the talents and capabilities of individual employees, who may leave and take their knowledge with them. In contrast, companies with a systematic and consistent system of documentation and decision-making develop a corporate body of knowledge which belongs to the organisation not to the individual. One only has to look at companies like McDonalds to see how this systems philosophy is used to achieve consistency of customer service and product quality across thousands of outlets with a relatively low-skilled, transitionary and low-paid labour force. Furthermore, in the highly competitive environment of modern business and the increasingly transitionary nature of employment relationships, those companies that fail to secure their corporate knowledge will quickly lose their competitive advantage.

MFM's solution was to develop the ROMS, which in itself would ensure greater consistency of practice in risk and opportunity management within MFM and along its supply chain. The ROMS ensures consistency of practice by:

- Providing a pre-selected list of practical tools, techniques and guidance for identifying, analysing and controlling risks and opportunities effectively. These were selected from the myriad available, after considerable debate to be carefully suited to MFM's unique customer base.

- Defining a system of reporting responsibilities so that everyone on an MFM project understands their own and others' roles in the risk and opportunity management process.

To enable a corporate body of knowledge to be accumulated in risk and opportunity management, the ROMS also contains a post-mortem process, which enables lessons to be learnt from the experiences of different managers in using the ROMS. A corporate knowledge-base in risk and opportunity management is accumulated by ensuring that any project-specific documentation from the ROMS is stored electronically on MFM's interactive intranet site. This information is analysed and disseminated regularly – an important feedback loop, which facilitates continual improvement in MFM's risk and opportunity management processes.

6.5.5.3   Complexity

Risk management is afflicted more than most areas of management, by complex terminology and jargon. In reflection of this, many companies tend to develop overly complex systems for managing risks, which intimidate people and give the impression that risk management is beyond their capability. However, risk management does not have to be complicated to be rigorous and it is important not to confuse rigour with complexity, as many companies do. It is far better to have simple and well-used procedures than complex ones that no one can understand and which are impossible to implement. Risk management is not rocket science. It is simply good management practice. If you impose a highly sophisticated system on an unprepared organisation, then you merely replace one set of risks with another set of risks which are more invisible and difficult to control.

MFM's solution was to ensure that the ROMS was written in plain English and was jargon-free, that it was user friendly and that there were a minimal number of forms to complete, which were simple and self-explanatory. This was important for Australian users but particularly important for non-English speakers on Australian and overseas projects. MFM also developed a unique four-tier system of complexity within the ROMS that allows users to choose a selection of risk management techniques, which are suited to their project's needs and their own capabilities. This flexibility is critical in encouraging people to use the system without feeling intimidated and it also provides a clear route for people to improve their risk management capabilities over time. It is also important in customising the ROMS to individual customer's needs. The four-tier approach to risk and opportunity management is explained in more detail, later in this case study.

6.5.5.4   Reactive rather than proactive

Many companies think they are practising risk management when they are actually practising crisis management. There is a tendency in the construction and facilities management industries to leave problems, let them grow and react to them when they have occurred, rather than dealing with them in advance. This is related to the common practice of separating rather than sharing risks, to the use of complex contracts which confuse risk management responsibilities, fragmented procurement supply chains and a general ignorance of proactive risk management practices. While crisis management skills in the construction and facilities management industries may have evolved to be second to none, this is nevertheless a very inefficient management strategy because problems which are neglected, inevitably grow in size and complexity and become much more difficult to resolve.

MFM addressed this problem by ensuring that decisions rather than problems triggered the risk and opportunity management process. The ROMS was designed to be triggered by the need to make a decision, rather than the need to resolve a problem or respond to an opportunity. In this way, the emphasis was very much on proactivity and encouraging people to think about their problems in advance. However, MFM recognised that it is impossible to predict all risks and opportunities in advance. Therefore, the ROMS also provided guidance and tools for reacting to risks and opportunities effectively. Reacting to problems effectively when they arise is critically important to ensure that they do not escalate into major crises. Conversely, reacting swiftly and effectively to opportunities when they arise is essential if they are to be maximised.

6.5.5.5   Offloading

Many companies rely on insurance and back-to-back contracts as a substitute for good risk management. There is a habit of passing responsibility for risk management down the procurement chain until it reaches the point of least resistance. By this time, the problem is usually a lot larger and more difficult to resolve. Many companies do not realise that transferring a risk inappropriately to a party which does not have the knowledge, capacity or capability to manage it, does not offload a risk. Rather, it merely gives the illusion of control and when the company that is given the risk is unable to deal with it, the seemingly offloaded risk will simply default back to the originator as a bigger problem that is much more difficult to manage.

MFM addressed this problem in a number of ways. First, the ROMS would incorporate guidance of good risk and opportunity management practice and deter managers from inappropriately passing manageable risks (and therefore potential opportunities) to their business partners. Second, by training staff in risk and opportunity management, they would better appreciate the dangers of arbitrary risk transfer and be more confident to deal with potential risks internally, turning them to advantage for MFM. Finally, MFM decided it would review all its risk transfer mechanisms (contracts, insurance strategies etc.) to ensure that they reflected this more confident and equitable approach to risk management.

6.5.5.6   Individualistic rather than cooperative

Typically, there is little collective responsibility for the management of risks and opportunities throughout the construction and facilities management supply chain. In the construction and facilities management industries the dominant culture is one of self-preservation and survival of the fittest – or more precisely, the more powerful. This selfish approach to the management of risk and opportunities is in no one's interests. A chain is only as strong as its weakest link and in the building procurement process, interdependencies are so strong that every one's interests are ultimately linked. There is little point managing one's own risks if one cannot manage those which arise along one's supply chain – from customers to suppliers.

MFM decided that the best way to achieve collective responsibility for risk management was to insist on the same standards and principles of risk management from their supply chain. They involved key stakeholders and business partners in the development of the ROMS, so that they understood what MFM wanted to achieve and what type of behaviour was expected from them. Also, the ROMS would need to incorporate extensive consultation arrangements, a requirement that everyone was expected to look out for and notify everyone else's risk and opportunities, as well as their own. Furthermore, MFM would require every party to an MFM contract to demonstrate compliance with the ROMS. In this way the ROMS would also drive supply chain reform. There was a belief that if MFM's business partners were not prepared to live up to MFM's high standards of risk management, then perhaps MFM should not be working with them at all.

6.5.5.7   Incestuous rather than consultative

Many of the risks and opportunities companies face and much of the information needed to manage them effectively can be found outside an organisation in its wider business environment and public community. A truly customer-focussed organisation recognises this and consults widely, meaningfully and appropriately with its stakeholders. Yet for an industry that has such a huge impact on society, the construction and facilities management industries are remarkably inward looking and insensitive to the needs of the many stakeholders affected by its activities. This incestuous culture results in many unnecessary problems. For example, it often ensures that companies do not understand customer's objectives as well as they could. It also results in a poor public image which makes it more difficult to raise venture capital and finance. Insurance is also relatively costly compared to other industries, and the construction sector in particular is relatively unattractive to the most talented youngsters and human resources in our society. Furthermore, the construction and facilities management industries all too often have to deal with frustrated government and regulatory bodies and an angry public protesting about the impact of its activities on their lives.

MFM solved this problem in a number of ways. First, it consciously decided to measure its success against its customers' objectives rather than against its own objectives. Few companies do this and the idea was that if MFM could keep its customers happy, then its business would automatically be successful, profitable and healthy. MFM also broadened its idea of who its customers were and ensured that it had extensive stakeholder consultation guidelines and requirements in the ROMS. The ROMS includes a simple stakeholder management tool which allows managers to identify and categorise stakeholders at the start of the risk and opportunity management process and consult them in an appropriate manner according to their interest in a decision outcome and their ability to influence it. MFM recognised that not only was it an increasingly stringent legislative imperative to consult stakeholders, but that badly managed stakeholders could be an enormous source of disruption and cost to its business. In contrast, well-managed stakeholders could be an enormous source of support, expertise and information which would help MFM manage its risks and opportunities more effectively.

6.5.5.8   Risk analysis rather than risk identification

There is a tendency for most risk management systems to focus on the process of risk analysis at the expense of risk identification. This is a function of the literature in risk management that very much focusses on the quantitative techniques of risk analysis. It is a significant problem because an unidentified risk or opportunity cannot be managed and, in many companies, simply getting employees to identify potential risks and opportunities effectively would be a major step forward.

The ROMS addresses this problem by requiring decision-makers to give as much attention to this stage as they do to any other stage of the risk management process. A whole section of the ROMS is dedicated to the process of risk identification. Indeed, this is the largest section within the ROMS and it provides a whole range of practical tools and techniques to ensure the effectiveness of the process.

6.5.5.9 Peripheral rather than core activity

In many companies, risk management is seen as a non-core, peripheral and even nuisance activity which has to be carried out to satisfy the mechanical requirements of system documentation. There is a widespread failure to recognise that the core activity of all businesses is risk and opportunity management – it is merely applied in different market contexts. It is therefore not surprising that risk management is rarely seen as a top management priority, is rarely championed and is rarely underpinned by a well-developed support infrastructure.

MFM responded to this problem by ensuring that senior management at director level promoted and championed the ROMS. Senior managers communicated risk and opportunity management as a core business function and led by example by rolling the ROMS out, from the top-down rather than from the bottom-up. Risk management activities are also being resourced properly by the gradual development of a comprehensive risk management support infrastructure including a training and feedback system, risk and opportunity database, specialist support staff and consultants and an interactive website. Effective HR support is also recognised as critical in providing training, performance appraisals and a reward system to recognise and promote good practice etc.

6.5.5.10 Centralised rather than decentralised

Most risk management systems see risk management as being under the responsibility of a centralised/senior risk manager. There are many problems with this approach. For example, many people tend to over rely on the centralised risk manager to manage their risks for them, but the risk manager is often too busy to manage the process effectively due to many other additional responsibilities. Most importantly, however, a centralised risk management system separates the responsibility for risk management from the point at which risks and opportunities arise, slowing down the response and separating it from those with the skill and experience to formulate the most appropriate response. Finally, a centralised risk management system depends too heavily on the skills, experience and management style of the risk manager.

MFM's solution was to decentralise the risk and opportunity management process, treating risk and opportunity management as the responsibility of all who are involved in making decision with significant risks and opportunities for MFM, its customers, its employees and business partners. The aim was to encourage people to take personal responsibility for their own decisions and not pass-the-buck down the line for someone else to solve. It was also hoped that this would push forward the responsibility for risk and opportunity management, to those who make the major decisions early in the procurement process and who create the majority of risks and opportunities in a project. In this type of decentralised system, the role of the risk manager becomes one of facilitation, championing, monitoring and controlling, rather than attempting, inevitably unsuccessfully, to manage everyone else's risk for them.

6.5.5.11   Periodic rather than continuous

Most risk management systems see risk management as a one-off event which occurs in risk review meetings at predetermined project milestones. The problem with this approach is that many risks and opportunities which arise between these milestones can be missed. Furthermore, when they are reviewed, the process is retrospective, meaning that risks can be enhanced and opportunities lost. Another problem, depending on who is involved in the risk review meetings, is the possibility of separating stakeholders from the response decision. When a large number of risks and opportunities have to be reviewed in these occasional meetings, the crammed agenda and limited time can make the risk review process a superficial one. The meetings can also be influenced too heavily by the expectations of senior managers who may not wish to hear bad news or even good news about potential opportunities if it involves extra work and time.

MFM's solution was to treat risk and opportunity management as a continuous process to be undertaken whenever a decision has to be made which could involve "significant" risks and opportunities for MFM, its customers, its employees and business partners. The aim was to encourage people to take personal responsibility for their own decisions and not defer them to a risk review meeting, which may not occur for several weeks.

6.5.5.12   Numerical

Unfortunately, society places considerable faith and value in numbers and if a decision-maker is able to support their argument with complex statistics, graphs and numbers then they are less likely to be challenged and more likely to be believed. There is also a perception, which arises from the actuarial origins of risk management, that if one is not using numbers then one cannot be doing risk management effectively. Not only does this dissuade the innumerate from practising risk management, it also encourages companies to over-rely on numbers to publicly legitimise decisions, causing them to ignore those risks which are less easily enumerated. Alternatively, there can be a tendency to attribute inaccurate numbers to risks that cannot be accurately measured and to ignore risks that cannot be easily enumerated.

In reality, while numbers may be very important for some areas of risk (particularly in commercial decisions relating to tendering, feasibility etc.), the majority of risks in the building procurement process cannot be easily enumerated. Indeed, even when they can be measured, many companies do not have the records and databases to do so reliably. Furthermore, as we said in Chapter 3, most employees may never fully understand probabilities and statistics. Finally, with the development and increasing popularity of PFI and PPP projects, it is increasingly the case that the majority of risks occur during the operational phases of a project rather than the early tendering and feasibility phases, where numbers are more relevant. It is a common and dangerous mistake to confine risk management to the early commercial stages of a project and to ignore the enormous operational risks which arise during the rest of its life – risks which often demand a far less numerical management approach.

MFM's solution was to ensure that there was a balance between quantitative and qualitative techniques in the ROMS. The important thing was to ensure that each would be used where appropriate, which meant that the system had to be flexible and adaptable to the needs of different situations and managers. In complex decisions, where managers have the necessary expertise and where quality data is available, quantitative techniques would be appropriate. In simple decisions not supported by reliable data where managers do not have numerical confidence, qualitative techniques would be used.

MFM also ensured that the use of the ROMS was not restricted to commercial decisions early on in a project. While the ROMS had to be suited to the management of commercial risks, it was imperative to ensure that the system would also be suitable for use during the operational phases of a contract. It is here that the majority of MFM's employees operate and, therefore, where the majority of risks and opportunities for MFM arise. MFM is most on view to its customers during the operational phase and it is here where reputational risks and opportunities which ensure repeat business are at their greatest.

6.5.5.13 Technology rather than people

Related to society's obsession with numbers is the tendency of many companies to invest in complex and expensive computer technology and risk modelling software. While it is very tempting to believe that all you have to do to manage your risks is to buy the latest computer program and press a few buttons on your keyboard, in reality no computer can yet replace the tacit knowledge that tells someone that something is not quite right. While computer technology can play an important part in the risk and opportunity management processes, it can also create a very dangerous illusion of control.

MFM's solution was to ensure that the ROMS was designed around people rather than technology. While MFM invested in the necessary technology to analyse high-end commercial risks and opportunities, it fundamentally believed that its employees were their most prolific source of information about risks and its best weapon for managing them. To this end, the ROMS was designed to illuminate the importance, talents and experience of high quality staff, encouraging an open flow of information between management and operatives, releasing the entrepreneurial spirit of employees and encouraging people to show initiative and creativity. The ROMS was seen as an important tool to enable MFM to capture, distil and effectively harness the collective knowledge and creativity of its workforce.

6.5.6   The end result

The final ROMS, is an unintimidating and self-contained document. Users are taken through the ROMS by a series of simple and incremental questions, which they would be likely to ask if they encounter the system or the concept of risk and opportunity management for the first time. To make the document appear user friendly, most questions are presented in the first-person such as "How do I know what types of decisions are covered by these guidelines?", "When do I start risk and opportunity management?", "What is my first step?", "What do I do when I have identified a risk?" etc.

Considerable thought also went into the design of the document, which was important from a psychological perspective. MFM did not want the ROMS to look intimidating and for this reason white space was maximised on each page, colour coding and images were used to enliven the document and help readers navigate, simplicity of approach was maintained by restricting the number of steps in any one stage of the process to no more than three, simple box diagrams were used to summarise processes and the margin on each page was used to summarise key points of advice.

The ROMS also includes real-life examples of how to use the various techniques of risk and opportunity management specified for use in the document. These techniques were carefully selected to suit MFM's business, from the many that have been developed in the risk management literature. However, it was clear that a manager on a simple residential facility would not have the same needs as a manager on a complex industrial facility. Similarly, a commercial manager would not have the same needs as a supervisor based on site. To cater for these different needs, the ROMS incorporates a flexible four-tier approach which allows a user to select a level of complexity which is suited to their capabilities and to the individual needs of their customer and contract. This is illustrated in Table 6.1. The appropriate level of operation in any situation is determined by a user's answers to a series of simple questions, which relate to the complexity of their business activity, the time they have available, the data that is available etc.

Table 6.1 ROMS' flexible four-tier system

Complexity	Level	Description
Simple	1	A minimum standard to be employed in all decisions that involve any *significant* risks and/or opportunities to MFM's employees, customers and business partners.
	2	A standard to be employed in decisions about business activities of *medium* risk and opportunity.
	3	A standard to be employed in decisions about business activities of *high* risk and opportunity.
Complex	4	A standard to be employed in decisions about business activities of *exceptionally high* risk and opportunity.

Risk management does not have to be complex to be effective and the benefits are not confined to large and complex projects or facilities. The important thing is that people make an appropriate choice and justify it. Of course, everyone in MFM is expected to do everything that is *reasonably practicable* to manage their risks and MFM's long-term aim is to enable all staff, through training, to be comfortable operating at a level which is appropriate to their responsibilities.

Having established the appropriate level of operation, the ROMS takes people logically, step-by-step, through the four stages of risk management, namely:

1 Risk and opportunity identification – this involves identifying *what*, *how* and *when* things can go wrong or improve.
2 Risk and opportunity analysis – this involves analysing the level of risk and opportunity by estimating the *likelihood* and *consequences* of potential events.
3 Risk and opportunity control – this involves deciding upon and implementing appropriate management responses.
4 Monitoring, reviewing and learning – this involves ensuring controls achieve their intended result and conducting project post-mortems to learn and communicate lessons for future projects.

As decision-makers work through the ROMS, the importance of the four-tier system becomes evident as they are taken on different paths through the document, the complexity of which depends on the level that is identified as appropriate to their needs. Users are referred to a series of pre-defined risk management techniques that correspond to their chosen level of complexity. For example, at Level 1 decision-makers are required to use simple techniques such as checklists and qualitative analysis and at Level 4 decision-makers use complex simulation techniques and quantitative analysis. In each case, the ROMS provides decision-makers with simple steps to follow and examples of how each technique is used in practice.

The Appendices to the guidelines are important and contain a glossary of terms, a standard checklist of common risks and opportunities, guidelines for consultation, lists of management responsibilities for risk and opportunity management at all levels of MFM and sources of potential information to assist people in managing their risks etc. They also contain standard forms for people to record their decisions and deliberations as they move through each stage of the risk and opportunity management process. The construction and facilities management industries are weak in good record keeping and documentation of decisions which are extremely important for auditing purposes. Every time a decision is made which uses the ROMS, these records are inserted into a single project-specific document called Risk and Opportunity Management Plan (ROMP). This is an extremely important document which starts its life in the inception stage of a project and is passed through each project stage until the project ends. This process ensures that the manager of each project stage knows that the risks and opportunities in previous stages have been managed effectively. Importantly, the ROMP can only be passed from one phase of the procurement process to the next once the previous phase ROMP has been signed off as having been satisfactorily completed. This process ensures that risks are not passed along the procurement chain and are dealt with at the point of creation and when they are most easily managed. The ROMP also provides a valuable audit-trail and basis for status-reports throughout a contract's life cycle and represents an important learning mechanism between different MFM contracts.

6.5.7   Implementing the ROMS

While the creation of the ROMS was a major undertaking, changing people's behaviour was far more challenging. No matter how well designed the ROMS was, people's acceptance of it would be critical to its success. The challenge was to establish the ROMS as an integral part of MFM's business culture so that it automatically determined the way that people operated, acted and behaved on a day-to-day basis. To minimise resistance to the implementation of the ROMS, Directors of MFM championed it and were intimately involved in its development. There was also extensive consultation with other MFM stakeholders, external and internal to MFM, at all stages of the system's development. This was achieved through a series of focus groups with senior MFM and Multiplex Construction staff who were responsible for assessing the effectiveness of the system, ensuring transferability across different business units and functions and putting a support infrastructure in place to ensure its successful implementation. External stakeholders such as customers, insurance providers, financiers, legislative bodies and other business partners were also consulted as part of this process.

A major result of this group's deliberations was a pilot study to test the ROMS on two key MFM contracts – Sydney's landmark and heritage-listed Queen Victoria shopping complex in the centre of Sydney and the Australian Defence Force's Maritime Base in Potts Point, Sydney. The lessons from this pilot study were enormously helpful in refining the system and in developing an effective implementation strategy. The main lessons from the pilot study are discussed below.

6.5.7.1   Consultation

The ROMS encouraged staff to consult more widely and more thoughtfully with stakeholders. As one of the contract staff said:

> The major benefit is that the system facilitates the need to communicate with the key stakeholders to ensure that we have identified the decision objectives and KPIs together. In other words we are not just analysing the decision requirements and solutions from our perceived key stakeholder requirements, we are confirming the same in a structured way and working on the decision as a team.

One of the benefits of better consultation was that surprising discoveries were sometimes made. For example, in one meeting, it was discovered that the contract team had misinterpreted the client's priorities, thinking that the client was primarily concerned with cost when its main concern was to meet a deadline. Consequently, it was also realised that contract performance measures could be improved to monitor the attainment of client objectives. New business opportunities also emerged from these stakeholder meetings. For example, in one contract, client pressure to execute additional projects in order to expend spend surplus monies in the client's budget always characterised the end of the financial year. This often caused problems for the client and MFM. However, as a result of the stakeholder consultation process, the contract team and client team agreed to work more closely to plan projects in advance and thereby bring forward client expenditures that would traditionally be left until the end of the financial year. Another benefit of these meetings was the goodwill generated as a result of improved communication with stakeholders, although not all stakeholders were used to being consulted in this way and needed to be educated about the process.

While there were many benefits generated from the consultation process, there were also some problems. For example, there was a general lack of confidence in running stakeholder meetings and staff felt that consultation could be dangerous if not managed effectively. There is clearly a skill involved in asking the right questions and preparing thoroughly for these meetings that needs to be learnt. Training was implemented to overcome this problem but external facilitation was considered useful where a wide variety of stakeholder interests had the potential to create conflict.

6.5.7.2   Negativity

As expected, people tended to focus on potential problems rather than opportunities. Although people were not used to formally thinking about risks, they were even less used to thinking about potential opportunities to do their job better than expected. To encourage more opportunistic behaviour, people were encouraged to consider potential opportunities *before* risks and an incentive system was instigated to enable them to share in any savings or improved performance they achieved.

6.5.7.3   Overenthusiasm

Some people in their enthusiasm to embrace the ROMS, threw themselves into the process and neglected to follow the systematic steps involved. It became clear that close monitoring of the ROMS implementation and simple training was going to be essential in avoiding this problem. A quarterly audit system was also developed to ensure that staff knew what exactly was expected of them and how their performance would be assessed. This is discussed in more detail later in this case study.

6.5.7.4   Time management

People were not used to putting dedicated time aside to think about risks and opportunities. Consequently, people tended to use the ROMS sporadically, fitting it around other duties, whenever they could find some spare time. In one project, the consequence of procrastination was that events overtook the staff, preventing new business opportunities being taken. In order to cope with the time demands of the ROMS process, a few people tried to delegate the process to a subordinate who did not have the skills, information and training to use it. Both strategies were equally dangerous. Close monitoring, training and regular auditing helped to avoid this problem, encouraging people to manage their ROMS time effectively to enable a focussed and continuous effort.

6.5.7.5   Learning curves

In using the ROMS, the learning curve for many staff was steep. Many had no previous instruction in risk and opportunity management and therefore were completely new to the idea and process. While younger staff were often relatively more informed about risk management than older staff, their lack of work experience ensured that they found the process much more difficult than experienced staff who could identify risks and opportunities more easily. To address this potential problem, staff were discouraged from working on the ROMS in isolation and encouraged to work in multidisciplinary groups with a spread of experiences.

6.5.7.6   Reactivity

The ROMS requires people to think proactively and in the long term, whereas client drivers are more often than not, short term and reactive. Understandably, people who had existed in this reactive environment for so long found it challenging to think proactively about risks and opportunities. Training and close monitoring was important in encouraging a proactive culture. Also, senior managers and customers needed to be educated about their role in the ROMS and the importance of being proactive in their planning and risk management activities. Generally, those at operational level are merely responding to the environment that has been created for them at strategic level.

6.5.7.7   Motivation

People needed to be motivated to use ROMS. The key barriers to motivation were identified as:

- Fear of change
- Fear of responsibility
- Fear of exposure
- Fear of learning
- Time constraints.

These were overcome by using positive rather than negative reinforcement. For example, MFM introduced an appraisal system which linked ROMS performance to staff performance bonuses. Risk and opportunity management responsibilities were also built into job descriptions and staff who demonstrated creativity, enthusiasm and imagination in using the ROMS were looked on more favourably in promotion rounds.

Other strategies to motivate people to use the ROMS included involving the pilot study participants in training, as champions to communicate its benefits. Last and not least, demonstrable commitment to the ROMS from senior management was critical as a motivational tool. There was some system fatigue in MFM when the ROMS was introduced and, if it was perceived to be just another system, then it would have failed. The system was therefore driven from the top-down, each organizational level helping the one below it to implement the ROMS effectively.

6.5.8 Training

Arguably, the biggest lesson from the pilot studies was the central importance of training to the effective implementation of the ROMS. For this reason, an innovative training strategy was developed, which was based on global trends in human resource development, the most significant of which is the shift from *training* to *performance improvement* (Philips 2001). The use of performance improvement programmes to replace traditional training programmes had a number of advantages for MFM:

1 Training and development would be linked directly to its strategic needs.
2 MFM would be forced to think carefully about what types of training and development programmes it needed and what tangible performance improvements they provide.
3 It would avoid any content that could not be directly linked to performance improvement, since performance improvement programmes would be based on practical and real business needs.
4 It would create the need to more closely monitor the relative costs and benefits of training and development programmes.
5 It would create the need to measure the return on investment in training and development programmes.
6 It would encourage line managers and HR managers to work more closely in establishing training needs.

In devising the ROMS performance improvement programme, MFM went through the process depicted in Figure 6.3.

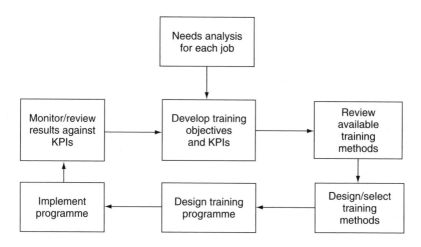

Figure 6.3 Steps in developing MFM's performance improvement programme.

Since the ultimate objective of the ROMS was to instil a risk and opportunity management culture into MFM, it was essential that the performance improvement programme would result in behavioural modification of the workforce. To this end, research indicated that it was essential that employees saw the programme as:

- Worthwhile – relevant to their needs, practical and value-adding
- Clear in its objectives
- Unintimidating
- Enjoyable
- Realistically achievable in terms of end results desired
- Supported by peers, colleagues and superiors
- Convenient – cognizant of existing work commitments
- Supported after the event by monitoring and feedback about implementation and performance – trainees must not be abandoned.

The final performance improvement programme was not a formal classroom-based programme but a practical, problem-based, self-directed, on-the-job learning programme, supported by a peer-based mentoring and coaching system and web-based learning. The process commenced with a volunteer being trained from each MFM contract to act as an ROMS champion, mentor and coach to other contract project managers/supervisors who collectively administered the ROMS on each contract. The volunteers liaised with the ROMS trainers to design and organise training suited to the needs and pressures of site-based staff. After this training was provided, their mentor role involved providing moral support and leadership while the coaching role involved practical advice and support in using the ROMS.

In addition to the above, there is a plan for all staff to be supported by an interactive website and by an email helpline. Beyond day-to-day email exchanges, there is a proposal for electronic component of the training strategy to be designed around a web-based training system known as WEBCT, which enables staff to study in their own time in an interactive medium. The system also allows staff to interact about their experiences via risk management noticeboards and to test themselves in using the system through randomly generated questions. The WEBCT system will automatically provide feedback on these tests and could eventually be used to automatically issue certificates to those who attained certain competence levels. This will allow senior managers to monitor the progress and performance of staff in improving their competency levels. In essence, e-learning is an interactive, flexible and cost-effective way for MFM to supplement and support the traditional training of its employees to use the ROMS effectively.

The performance improvement programme was designed to change MFM's risk management culture by evolution rather than revolution. To this end, rather than training the whole workforce at once, the ROMS

Figure 6.4 Stages in ROMS training process.

training was rolled out gradually as new contracts came on stream, starting in the commercial planning/feasibility/tendering phase, moving into the transition phase and ending in the operational phase of the building's life cycle (see Figure 6.4). This top-down and gradual approach ensured that senior management would be leading and championing the ROMS process.

To ensure that the MFM supply chain also managed its risks and opportunities effectively, MFM is currently developing a programme whereby all sub-contractors, suppliers and consultants on MFM contracts will undergo a one-hour ROMS induction process. The purpose of the induction is to inform them that the ROMS exists, to clarify expectations and answer any questions about the ROMS. They will not be expected to use the ROMS, but will be expected to comply with requests for information from MFM staff in conducting their ROMS responsibilities and to assist in any way they can in this process. Business partners will also be expected to have equivalent quality risk management systems which will be assessed in a revised pre-qualification process.

6.5.9 ROMS audits

To support the training programme, quarterly audits are conducted to provide positive reinforcement and feedback to contract staff on their use of the ROMS. The audits also provide MFM with important information about the ROMS's impact on business performance.

6.5.10 Support structure

MFM developed a range of strategies to support those who would be using the ROMS.

First, this involved the formation of a risk management team with specific responsibilities for implementing, managing, monitoring and updating the guidelines. The structure of this team is depicted in Figure 6.5 where the wider resource implications of developing a risk management system become clear.

MFM is also in the process of establishing a risk and opportunity management section on its existing intranet site which will allow those working on MFM facilities to share, visualise and communicate information about risks with clients, staff, sub-contractors, suppliers, consultants and authorities, anywhere in the world. On this intranet site staff will be able to find:

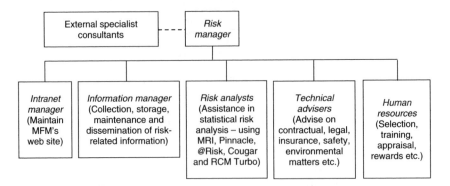

Figure 6.5 ROMS support structure.

- An electronic copy of the ROMS guidelines
- Standard forms which can be downloaded for use
- A dedicated facility-specific ROMP template
- A database which provides information to help you manage risks effectively
- A library of useful references
- A risk and opportunity management noticeboard that is continually updated with useful tips, contacts and information
- A discussion page that will enable managers to share their experiences
- A suggestion box that allows managers to provide ideas for improvement
- A helpline where you can consult an expert by email for further information.

6.5.11   The benefits of the system

The benefits for MFM of developing and implementing the ROMS have been numerous. For example, it has been a strong and visible statement of MFM's corporate values, communicating clear expectations of employee behaviour and intended public image. It has also simplified MFM's existing risk management practices and achieved greater levels of consistency across different MFM systems (Occupational Health and Safety, Environmental Management, Quality Management). The ROMS has also allowed MFM to demonstrate that it is a good corporate citizen and that it has a strong commitment to its corporate, legal, moral and ethical responsibilities to its customers and to the community as a whole. Finally, MFM have already begun to identify opportunities to improve performance beyond customers' expectations – opportunities that may have gone undetected before the ROMS was developed. Ultimately, the new system will result in better quality documentation for decision-making, improved record keeping, fewer unforseen problems and greater opportunities. In particular, the ROMP, which is the end result of the ROMS process provides an excellent record and audit trail of decision-making throughout the life of a contract, providing a basis for status reports and evidence that MFM has done all that is reasonably practicable to mitigate its risks and maximise it opportunities. The ROMP is also an excellent learning mechanism between projects, allowing MFM to continually improve its processes over time.

In the longer term, MFM are confident that the ROMS will result in lower insurance premiums, cheaper finance and more open, transparent and trusting relationships with its customers. Much of this will depend on MFM demonstrating that the ROMS makes a meaningful and enduring difference to the quality, consistency and reliability of its services. Ultimately, the long-term effectiveness of the ROMS will be measured by the degree that it has changed people's attitudes and working practices in MFM and how it has improved the management of MFM's customers' facilities. This is the most important measure of a truly customer-focussed system.

The overall benefits of developing and implementing the ROMS can be seen in Figure 6.5 which compares the result of a risk management maturity audit before and after the introduction of the ROMS. Figure 6.6 illustrates that procedures for managing risks and opportunities have been dramatically improved in MFM, as has the overall awareness of risk management throughout the organisation. Extra resources have also been dedicated to risk and opportunity management and, as the company continues to grow, it will be possible to allocate more dedicated resources. Skill and confidence levels have also begun to improve as a result of the training programme and will continue to do so as MFM staff gains more experience in using the ROMS over time. The culture of MFM has also begun to change as people

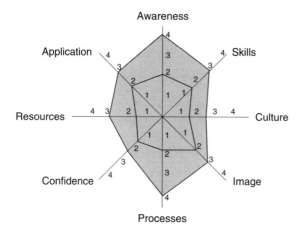

Figure 6.6 Changes in MFM after introducing the ROMS.

start to realise that risk and opportunity management should be an automatic way of thinking whenever decisions are made. However, changing the culture of MFM will inevitably take time, patience and perseverance. The monitoring and positive auditing process will play a key role in this process.

6.5.12   Conclusion

In theory, the perfect risk management system is deceptively simple to produce. However, in reality the task of developing and implementing an effective system is a demanding and often frustrating process, which requires vision, leadership, and a considerable amount of patience, time, resources, commitment and determination. The ideal risk management system is a continually moving target and it is unlikely that you will get it right first time. Furthermore, there will inevitably be barriers in your way and, in many instances, the resistance to change will come from unexpected quarters. There is a great degree of politics to manage in introducing a system of this magnitude and there will be plenty of opponents who believe that the organisation is doing just fine as it is. Strong leadership is critical in overcoming these organisational defences.

One of the main lessons that have been learnt in this project is that developing a ROMS is relatively easy compared to implementing it. The implementation process took far more time and resources than first thought. Ultimately, the success of any system depends as much on people changing their behaviour as it does on a well-designed system and achieving this is not an easy task. One way of avoiding problems in the initial phases of implementation of a system is not to set expectations of behavioural changes too high. It is critically important to be realistic in what can be achieved and to set a realistic timetable for implementation. In the case of the ROMS, the timetable for full implementation was five years. Initially, the aim was simply to get people rigorously identifying their risks, in the hope that once they had been identified, most managers would want to analyse them further. A second important point is to ensure that risk and opportunity management is not seen as something extra to be done on top of normal activities. The greatest chance of acceptance is when risk and opportunity management is seen as a part of normal business processes. For example, most organisations produce an annual business plan of which risk and opportunity management should be a normal component. By pointing this out, risk and opportunity management can be portrayed as a help rather than a hindrance.

We have said a number of times, that introducing risk management into an organisation involves cultural change. We have also said that to expect people to start thinking proactively when all of their traditional drivers are short-term and reactive, will inevitably be a major challenge, especially if the traditional drivers are left in place. Consequently, it is important to tackle both the symptoms and causes of this problem and to educate your clients and supply chain about their role in changing the culture. Furthermore it is critical to have a continuous monitoring system in place for a considerable time after introducing the system, otherwise people will revert back to their old behaviour. One should not underestimate the importance of an effective human resource management function in bringing about cultural change. An effective training system, motivation and performance appraisal system play a critical role in bringing about behavioural change and keeping it changed.

Finally, it is important to realise that developing a risk management culture is something which cannot occur overnight. It is a journey which takes time and commitment over an extended period. Therefore, it is essential that MFM continue to invest in risk management processes, championing the ROMS and monitoring its implementation on a continual basis, learning lessons, feeding them back into the company and refining risk and opportunity management practices. Every means possible must be used to develop and maintain a strong and pervasive risk management culture and to ensure that all new business partners and employees are educated about the ROMS and MFM's high expectations of practice. In the longer term, MFM will need to identify and counteract any evidence of risk management fatigue by publicising and celebrating good practice and improvement statistics, rewarding those who excel and using regular relaunches and promotions to renew the ROMS process. Maintaining training to update and reinforce risk management skills is also essential. A loss of momentum could easily derail the significant advances which have been made in MFM, and ultimately lead to a loss of credibility for the ROMS. Of particular importance is the need to build a database of risk and opportunity information and statistics to provide the basis for more complex risk and opportunity management in the future. Ultimately, the quality of risk management, like any management process, is completely dependent upon the quality of the data it is based on. To make this manageable, MFM have decided to initially focus on their main areas of risk and opportunity and then later expand their database into other areas.

Appendix A

Risk management maturity level checklist

	Level 1 Ad hoc	Level 2 Established	Level 3 Managed	Level 4 Integrated
Culture	No risk awareness. Risk management seen as a nuisance and peripheral activity with no relevance or value to core business objectives. No upper management involvement or support. Resistance and reluctance to adopt risk management. Tendency to continue with existing processes even in the face of project failure.	Risk processes are viewed as a compliance requirement *and* an additional overhead with variable practical benefits. Scepticism of ability of risk management to add value to organisation. Focus on downside of risk. Risk management system is primarily for public relations purposes but not implemented. Upper management encourages but does not require risk management.	Benefits of risk management recognised, accepted *and* proven. Focus on upside *and* downside of risk. Upper management requires risk reporting. Bad news risk information is accepted. Informal communication channels to top management.	Risk management widely seen as a core business function. Risk is an instinctive and automatic way of thinking for all employees at all levels of organization. Open flows of information and trusting relationships with business partners along entire supply chain. Collective responsibility for risks and opportunities along supply chain. No blame culture – acceptance of mistakes.

	Managers do not want to hear about problems. Many undiscussable problems. People are punished for communicating bad news. Secretive inward looking – no stakeholder communication.	Little communication with stakeholders.	Effective communication with stakeholders.	...stake... actively encouraged through formal mechanisms to participate in business decisions.
Processes	No structured and documented approach to deal with risk. No formal processes. No risk management plan. Reactive management of risks. Over reliance on insurance as a substitute for effective risk management. A policy of risk transfer to weaker parties through contractual mechanisms. Internal business processes actively create risks.	Project-based risk management systems with little inter-relationships. No generic processes. No risk management planning across projects. Risk management processes inconsistent across different management systems. No attention to reducing risk exposure created by internal business processes.	Generic risk management processes widely communicated and implemented on most projects and common across different management systems. Risk metrics collected to support basic quantitative analysis. A policy of risk fairness in contracts rather than risk transfer. Steps activity taken to reduce risk in products, services, business processes and production processes. Use of external experts and services in risk management.	Risk-based organisational processes at all levels and functions of organisation. Well-developed, tested and refined risk management procedures. Regular monitoring, evaluation, auditing and improvement of processes. Management of risk built into all organisational processes and consistent across all management systems. Wide range of reliable risk metrics covering whole risk portfolio collected and analysed systematically. Processes reflect good principles of risk management/ transfer – re; pricing, capability, resources must be appropriate to risk.

	Level 1 Ad hoc	Level 2 Established	Level 3 Managed	Level 4 Integrated
				Diversification and portfolio strategies in place. Computerised inventories of plant, employees, products and capabilities. Business continuity planning, crisis management and emergency systems in place and regularly tested – backed up by technical redundancy. Regular legal and financial audits of threats and opportunities undertaken. Dedicated research on hidden opportunities and threats. Critical follow up and learning from incidents.
Awareness	Unaware of the need for risk management. Little or no attempt to learn from past projects.	Experimenting with risk management through a small number of enthusiastic individuals. Aware of potential benefits of managing risk but no effective implementation. Staff tends to react to risks as and when they arise.	Benefits of risk management understood at all organisational levels and along supply chain, although not consistently. Key internal stakeholders and suppliers can participate in risk management process. Proactive approach to risk when making decisions.	Risk awareness applied proactively in making all decisions. Risk awareness instilled throughout all organisational levels and along entire supply chain. Active use of risk feedback to improve organisational processes and gain competitive advantage. Collective responsibility for risk along entire supply chain. Key suppliers, external and internal stakeholders and customers participate in risk management process.

Skills/ Experience	No understanding of risk management language or principles.	Basic understanding of risk management language or principles in organisational pockets. Limited to individuals who have had little or no formal training. No analysis capability apart from some basic qualitative analysis by individual managers.	Widespread understanding of risk management language or principles. Qualitative analysis is widely practices and some basic quantitative analysis.	Intimate and developing understanding of risk management language or principles and how it applies to organisation's business. Where appropriate, complex quantitative analysis is possible using sophisticated probabilistic and simulation techniques. State of the art tools and methods in use. Evolving corporate memory of and learning about past risks and opportunities.
Image	Reputation for poor risk management associated with cost overruns, delays, poor safety, poor quality on projects.	Perception of competence but unreliability associated with variable performance and well publicised problems on contracts spreading between clients.	Reputation for effective risk management consistency of service, and product quality based on well publicised and widely implemented risk management system.	Reputation for excellent risk management acquired from successful completion of high-risk projects. Company attracts educated clients which are sophisticated in risk management and expect same standards. Customers have confidence that organisation can take on higher risks than competitors. Added value to customers often added by emphasis on upside as well as down side of risk. Major efforts in public relations and stakeholder management.

(Continued)

	Level 1 Ad hoc	Level 2 Established	Level 3 Managed	Level 4 Integrated
Application	No or very few managers practice risk management.	Risk management applied inconsistently in response to customer demands. Risk management inconsistently practiced on selected projects depending on knowledge of managers on those projects.	Risk management applied consistently across systems and levels but needs continuous support and leadership to maintain. Risk management focused on operational risks. Risk management training.	Risk management consistently and systematically implemented on all projects and across all management systems. Enthusiasm for value of system develops its own momentum for continuous improvement. Risk management applied to broad range of risks – political, reputational, strategic, commercial, and operational. Regular risk management training to all staff.
Confidence	Fear of risk management. No experience in implementing risk procedures. No confidence in identifying, analysing and controlling risks.	Fear of risk management remains in pockets. Risk analysis beyond most people – better risk identification processes are a major step forward.	Perceptions of fear have been broken. People work confidently at own ability level and actively seek further information to help manage risks. Support system in place to help people with risk management activities.	Overt confidence in managing risks communicated to customers and clients. Enthusiasm to learn about risk management and develop skills. Staff see risk management as their core skill. Interactive and intelligent support system available to staff which enables learning across different functions.

				Risk management system develops a life of its own – driven forward and developed by staff. Risk leadership provided by staff. Staff externally communicate risk management capabilities as a competitive advantage.
Resources	No dedicated resources for risk management.	All risk personnel located under project. No central support. Risk financed under project cost centres.	Top management commitment to risk management. Active allocation and management of risk budgets. In-house core of expertise, formally trained in basic risk management skills. Development and use of specific dedicated processes and tools for business. Training of key people in organisation who administer and involved in risk management system.	Dedicated budget/ resources for risk management. Top-down implementation of system led by strong management leadership. Dedicated risk management unit or team. Centralised risk management expertise and resources and support for everyone in the organisation. HRM support risk management activities through incentives, training, rewards etc. Resources to support, train supply chain in risk management. Psychological support for employees, stress management,

Source: Adapted from PMI, 2002; Mitroff and Pearson, 1993; and Loosemore, 2000.

A checklist of common risks and opportunities in construction projects

Technology
- Availability of materials/technologies/equipment.
- Experience of working with materials/technologies/equipment.
- Lead times for orders of materials/technologies/equipment.
- Stability of design, design changes etc.
- Availability of key components and spares.
- Equipment reliability/safety/productivity.
- Innovation – need for further development.
- Maintenance and spare parts costs.
- Reliability, maintainability, availability, support availability.
- Specification completeness and accuracy.
- Clarity of technical performance, standards or regulations.
- Technological change, updates, obsolescence.
- Materials quality/safety.
- Workmanship.
- Productivity of equipment.
- Availability of critical plant/equipment/spare parts, fuel, skills for operating etc.
- Sampling/testing.
- Ground conditions (mining activities, rock, services, antiquities, contamination etc.).
- Suitability, availability and supply of materials.
- Specialist equipment (knowledge of, skills, training, difficulty of use, consistency of use, cost etc).
- Transport (difficulties, availability, suitability, police and liaison requirements, usage constraints, site access, noise, pollution, weather impact etc.).
- Control over design process (opportunities for influence over design decisions, designers' understanding of issues, communication with designers etc).
- Availability of design information/design changes.
- Quality of design (buildability, omissions, incompatibility between different designs, details, components, sub-standard performance when built, difficult to build etc).
- Innovation in design (level of standardisation, newness of technology/details/materials etc).

	• New technology (unfamiliarity, application, feasibility, specialist controls/monitoring needed).
	• Software (theft, misuse, abuse, database size and complexity, development required, training required etc).
	• Security.
	• Product contamination.
	• Product safety, safety guidelines, hazardous materials etc.
	• Overseas voltage compatibility.
Human	• Effectiveness of communications (language difficulties, use of translators, accuracy of translators etc.).
	• Working and living conditions for staff.
	• Crimes against people, property, vandalism, bribery, espionage, terrorism and extortion.
	• Security and safety of staff, personnel and public.
	• Industrial relations.
	• Labour sickness/absenteeism.
	• Quality, capability, reliability, productivity and availability of labour (operatives (sub-contractors) and managers)
	• Attitudes of staff towards quality, costs, environment, safety, trust, opportunities etc.
	• Staff reliability, skills, capability etc.
	• Culture (compatibility, different ways of working, different standards, different priorities, cultural assimilation etc).
	• Personality conflicts.
	• Skills and staffing issues (Adequate prior experience, availability and mix of skills staff, learning curve effects, loss of critical skills/staff, staff turnover, recruitment, induction, training needs/timeframe/effectiveness, willingness of key staff to relocate etc).
	• Intimidation/racism/discrimination.
	• Malicious damage/sabotage to property/vandalism.
	• Theft.
	• Bribery.
	• Corruption.
	• Malicious attacks upon individuals/personal conflicts.
	• Sabotage.
	• Mistakes/errors/incompetence.
	• Stupidity.
	• Inefficiency.
	• Personality conflicts.
	• Negligence.
	• Differing professional/personal values and beliefs.
	• Different ways/methods of working.
	• Interference between trades.
	• Communication effectiveness.
	• Misunderstandings/misinterpretation.
	• Cultural differences (language, traditions, food, beliefs, religious etc).
	• Indecisiveness.
	• Unreasonableness.

Environmental	• Force Majeure (acts of god) – heat-wave, rain, wind, heat, cold, humidity, fire, tidal wave, volcanic, earthquake, flood, storms/cyclones, landslide, lightening strike etc.
	• Pest/vermin infestation.
	• Industrial/environmental disaster.
	• Pestilence.
	• Disease and health risks.
	• Pollution.
	• Ecological damage.
	• Endangered species.
	• Contamination of land, water and air.
	• Conservation.
	• Hazardous chemical or gas release.
	• Hazardous sites and materials.
	• Legislative and regulatory constraints
	• Noise.
	• Waste, recycling etc.
Commercial and legal relationships	• Overseas (compatible or familiar legal system, corruption).
	• Finance.
	• Conditions of contract (clarity, complexity, fairness, variations to standards).
	• Conditions of acceptance.
	• Purchase agreements.
	• Contractual requirements (joint venture partners, local labour etc).
	• Disputes/claims.
	• Breaches of contract (liquidated damages etc.).
	• Completion dates.
	• Potential professional liability.
	• Fraud/corruption.
	• Enforceability of contracts with governments.
	• Delay in possession of site – access problems.
	• Determination.
	• Contract revisions/changes.
	• Supply (shipping/delivery – materials, plant and labour).
	• Delay in resolving disputes.
	• Costs of obtaining decisions.
	• Delays in enforcing decisions.
	• Joint venture partner (potential for litigation, complexity of contracts and documentation, level of control and responsibility, complexity of organisational structure etc.).
	• Overseas contracts – approval processes, level of prescription, local conditions, local biases, corruption, corporate and licensing arrangements/regulations etc.
Management activities/controls/systems	• Documentation quality (errors/omissions in bill of quantities, inadequate information, conflicting information on designs, poor specification, unbuildable designs, inaccurate estimates etc).

- Information to make decisions (poor databases, out-of-date information, late information, inaccurate information, ambiguous information, unusable information, difficult to understand information etc).
- ⊙ Quality of planning – realistic programmes etc.
- Coordination between different sub-contractors.
- Motivation.
- Communication.
- Decision-making.
- ⊙ Leadership.
- Crisis preparedness – emergency plans.
- ⊙ Level of planning (fast-tracking/design completion).
- ⊙ Monitoring/control systems (quality, costs, time).

Economic/financial
- ⊙ Availability of funds/funding sources.
- Joint venture partners (requirements for skills/equity).
- ⊙ Complexity of funding arrangements (equity funding and ownership).
- Financing costs/terms/conditions – interest rates, investment conditions etc.
- ⊙ Availability of extra funds if needed.
- Agents fees.
- Price fluctuations/inflation.
- Market factors/revenue fluctuations.
- Market competition.
- Demand change.
- Demographic changes.
- Exchange rate fluctuations.
- Currency value changes.
- Funding/payment problems/constraints.
- Cash flow problems (us, sub-contractors, suppliers, client).
- Bankruptcy/insolvency (Ditto).
- Payment problems/delays/valuation.
- Pay demands/constraints.
- Local and national taxes (payment, avoidance, enforcement).
- Cash management/foreign exchange management.
- Working capital requirements.
- Market factors influencing cost of working capital and prices of goods and services offered and bought (e.g. energy, materials, maintenance).

Business partners (principals, partners, contractors, sub-contractors and suppliers)
- ⊙ Communication (proximity, remoteness, infrastructure, systems).
- Liquidity, solvency, business failure, stability, change of ownership, vulnerability.
- Ability and willingness to pay, payment philosophy etc.
- Credit rating.
- ⊙ Client/user/operator – clarity and compatibility of expectations/perceptions/requirements.
- Ability to meet contract commitments.
- ⊙ Attitude to changes in scope, specifications, costs, time etc.
- Attitude to litigation.

- Availability of information about business.
- Attitudes towards environment, safety, quality, time, cost etc.
- Ability to take delivery of product, transition arrangements etc.
- Principal interaction.
- Representatives – personalities, number, experience etc.
- Trust.
- Past relationships.
- Future business relationship.
- Technical knowledge, skills and experience to deal with specified technologies.
- Availability.
- Compatibility (business culture, systems, geographic, personalities, assimilation requirements etc.).
- Speed of response, bureaucracy, decision-making, communications with etc.
- Organisational structure (general and for contract).
- Ability to deliver on time and within budget.
- Lead times for orders.
- Accreditation for safety, quality etc.
- Alternative suppliers/sub-contractors etc in case of insolvency, failure to deliver etc.
- Commercial terms.
- Contract team, management, control and supervision.
- Costs of goods and services, costs of extras, costs of delays (consequences if things go wrong) etc.
- Warranty on goods and services.
- Quality of goods and services.
- Reliability of goods and services.
- Failure in supplier/sub-contractor supply chain.
- Quality of supplier/sub-contractor supply chain.
- Flow-on conditions, criticality of service, dependence upon.
- New or existing contract/supplier etc.
- Joint venture partner (financial stability, compatibility, technical resources, skills, knowledge, expertise, commitment, familiarity).
- Overseas (quality of business agents).

Political

- Customs/export/import restrictions (embargos etc).
- Overseas (political/social stability, cultural and religious factors, relationship with host government, international relations, support from embassy/trade commission, prejudice against foreign companies/employees, lobbying etc).
- Requirements for permits etc.
- Ability to repatriate profits.
- Availability of foreign exchange.
- Changes in government.
- Changes in government policy.
- Changes to taxation/royalty regimes.
- New legislation (taxes, labour, safety, waste, environmental etc).
- Expropriation.

- Government constraints on operations.
- ⊙ Government endorsement/intervention.
- Native title, native owners issues.
- Import/export restrictions.
- Withdrawal of approvals and licences.
- Enforcement bodies (inspections relating to legislation).
- ⊙ Local government/services (liaison, planning, approvals, inspections etc).
- Civil commotion, wars.
- ⊙ Public relations/perceptions.
- ⊙ Pressure groups/canvassing/consultation/protests/support (environmental, disabled etc).
- ⊙ Public reaction/complaints/perceptions/misperceptions/ protests (relating to dust, noise, lighting, security, inconvenience etc).
- Probity.

References

Adams, J. (1995) *Risk*, UCL Press, London.

Akintoye, A., Beck, Hardcastle, C., Chinyio, E. and Assenova, D. (2001) *Framework for Risk Assessment and Management of Private Finance Initiative Projects*, Glasgow Caledonian University, Glasgow, Scotland, UK.

Amabile, T. M. (1983) *The Psychology of Creativity*, Springer-Verlag, New York.

Anonymous (2003) Planning for a rainy day, *Property Australia*, 17 (9), 45–49.

Argyris, C. (1990) *Overcoming Organisational Defenses*, Allyn and Bacon, London, UK.

Arndt, R. and Maguire, G. (1999) *Private Provision of Public Infrastructure: Risk Identification and Allocation Project*, survey report, Department of Treasury and Finance, Melbourne, Victoria.

AS/NZS 4360: 1999 *Risk Management*, Standards Australia, Canberra, Australia.

ASCE (2002) TISP Congress addresses challenges in protecting the built environment, *ASCE News*, December, 3–4.

Asch, S. E. (1956) Studies of independence and conformity: A minority of one against a unanimous majority, *Psychological Monographs: General and Applied*, 70 (416), 1–70.

Aspery, J. (1993) The media: Friend or foe? *Administrator*, 2 (2), 17–19.

Aspery, J. and Woodhouse, N. (1992) Strategies for survival, *Management Services*, 36 (11), 14–16.

Barnes, M. (1991) Risk sharing in contracts. In *Civil Engineering Project Procedure in the EC*. Proceedings of the conference organized by the Institution of Civil Engineers, Heathrow, London, January, 24–25.

Barnes, P. (2002) Approaches to community safety; risk perception and social meaning, *Australian Journal of Emergency Management*, 15 (3), 15–23.

Basadur, M., Graen, G. B. and Scadura, T. A. (1986) Training effects on attitudes toward divergent thinking among manufacturing engineers, *Journal of Applied Psychology*, 71, 612–617.

Bea, R. G. (1994) *The Role of Human Error in Design, Construction, and Reliability of Marine Structures*, Ship Structure Committee SSC-378, U.S. Coastguard, Washington, DC.

Beck, U. (1992) *The Risk Society: Towards a New Modernity*, Sage Publications, Newbury Park, CA.

Behling, O. and Eckel, N. L. (1991) Making sense out of intuition, *Academy of Management Executive*, February, 46–47.

Belbin, R. M. (1997) Conventional wisdom, *People Management*, 3 (1), 36–38.

Bernstein, P. L. (1996) The new religion of risk management, *Harvard Business Review*, March-April, 74 (2), 47–52.

Berry, A. J. (2000) Leadership in a new millennium: The challenge of the "risk society", *The Leadership and Organisation Development Journal*, 21 (1), 5–12.

Bingham, T. (2000) Shoes made for walking, *Building*, June, 53.

Blockley, D. (1996) Process re-engineering for safety. In James, M. (ed.) *Risk Management in Civil, Mechanical and Structural Engineering*, The Institution of Civil Engineers, Thomas Telford, London.

Bookstaber, R. (1999) Risk management in complex organisations, *Association of Investment Management and Research*, New York, 18–20.

Boothroyd, C. and Emmett, J. (1996) *Risk Management – A Practical Guide for Construction Professionals*, Witherby Publishers, London.

Bowden, A. R., Lane, M. R. and Martin, J. H. (2001) *Triple Bottom Line Risk Management*, John Wiley and Sons, New York, USA.

Boyd, H. (1990) Case study: The Zeebrugge car Ferry Disaster. In Frederick, W. and Post, J. (eds) *Business and Society: Case Studies in Corporate Social Policy*, McGraw Hill, Singapore, 498.

Brown, R. (1998) Tales of the unexpected, *Works Management*, 51 (11), 63–65.

Burby, R. J. (2001) Involving citizens in hazard mitigation planning: Making the right choices, *Australian Journal of Emergency Management*, Spring, 45–58.

Burns, T. and Stalker, G. M. (1961) *The Management of Innovation*, Tavistock, London.

Button, R. (2002) Do I look like a terrorist? *RICS Business*, September, 7–8.

Cameron K. S. (1984) The effectiveness of ineffectiveness. In Staw, B. M. and Cummings, L. L. (eds) *Research in Organisational Behaviour*, Vol. 6, JAI Press Inc., Greenwich, Connecticut, 235–285.

CDM (1994) *Construction (Design and Management) Regulations* (1994) HMSO, London.

Chapman, R. J. (1998) The effectiveness of working group risk identification and assessment techniques, *International Journal of Project Management*, 16 (6), 333–343.

Chapman, C. and Ward, S. (2002) Project risk management; the required transformations to become project uncertainty management. In Slevin, D., Cleland, D. and Into, J. (eds) *The Frontiers of Project Management Research*, Project Management Institute, Pennsylvania, 405–417.

Chappell, D. (2001) *Parris's Standard Form of Building Contract*, Third Edition, Blackwell, Oxford, UK.

CIS (2002) Congress on infrastructure and security for the built environment, First congress, The infrastructure security partnership (TISP), 5–7 November, Washington, DC.

Cook, A. (1999) Specialists call on state to stop withholding cash, *Building*, 5 November, 13.

Cottle, G. (2003) Risks associated with public private partnership arrangements and private finance initiatives, *Through the Looking Glass*, FMA Ideaction Conference, 7–9 May, Sydney, Australia.

Davenport, P. (2000) *Adjudication in the NSW Construction Industry*, The Federation Press, Sydney, Australia.

Department of Housing and Construction (DHC) (1980) *Life Cycle Costing*, Technical Information Bulletin T1140 AE, Department of Housing and Construction, Sydney, Australia.

Engineering Council (1999) *Guidelines on Risk Issues*, Engineering Council, London, UK.

Edkins, A. and Millan, G. (2003) *Construction Risk Identification: A Review of the Literature on Cognitive Understanding of Risk Perception*, The Bartlett School of Architecture, Building, Environmental Design and Planning, Paper number 16, University College London, London, UK.

Edwards, L. (1995) *Practical Risk Management in the Construction Industry*, Thomas Telford, London.

Fenton-Jones, M. (2003) Intellectual theft a rowing problem for business, *Australian Financial Review*, Tuesday, 21 October, 49.

Fischhoff, B. (1995) Risk perceptions and communications unplugged: Twenty years of progress, *Risk Analysis*, 15 (2), 137–145.

Flanagan, R., Kendell, A., Norman, G. and Robinson, G. D. (1987) Life cycle costing and risk management, *Construction Management and Economics*, 5 (10), 53–71.

FMA (2004) *Facility Management Guidelines to Managing Risk*, Facility Management Association of Australia Limited, Melbourne, Australia.

Frazer, N. M. and Hippel, K. W. (1996) *Conflict Analysis: Models and Resolutions*, North Holland, New York, USA.

Freeman, R. E. (1984) *Strategic Management: A Stakeholder Approach*, Pitman, Boston, USA.

Friedman, S. (2000) There's more than one way to skin a "cat", *National Underwriter*, Chicago, May, 104 (19), 260–263.

Furmston, M. (2003) *Construction Contracts*, Blackwell, Oxford, UK.

Furze, D. and Gale, C. (1996) *Interpreting Management – Exploring Change and Complexity*, International Thompson Business Press, London, UK.

Ginn, R. D. (1989) *Continuity Planning: Preventing, Surviving and Recovering from Disaster*, Elsevier Science Publishers Ltd, Oxford, UK.

Glackin, M. and Barrie, G. (1998) Jubilee Line strike set to end after last-ditch talks, *Building*, 27 November, 12.

Gomez-Mejia, L. R., Melbourne, T. M. and Wiseman, R. M. (2000) The role of risk sharing and risk taking under gain-sharing, *The Academy of Management Review*, Mississippi State, 25 July, 3, 492–507.

Greenspan, A. (1999) Measuring financial risk in the twenty first century, *Vital Speeches of the Day*, New York, November, 66 (2), 34–36.

Grimsey, D. and Lewis, M. (2004) Discount debates: Rates, risk uncertainty and value for money in PPPs, *Public Infrastructure Bulletin*, March, 4–7.

Gulliver, F. R. (1991) Post-project appraisals pay, *Harvard Business Review*, Paperback Number 90053, 65–68.

Hartman, F. T., Snelgrove, P. and Ashrafi, R. (1998) Appropriate risk allocation in lump sum contracts – who should take the risks, *Cost Engineering*, Morgantown, July, 40 (7), 21–26.

Hemsley, A. (2002) Prepare for the high jump, *Building*, 10 May, 52.

Hersch, J. (1998) Smoking and other risky behaviours, *Journal of Drug Issues*, Tallahassee, Summer, 28 (3), 645–661.

Hillson, D. (2002) Extending the risk process to manage opportunities, *International Journal of Project Management*, 20 (2), 235–246.

Hood, C. and Jones, D. K. C. (1996) (eds) *Accident and Design – Contemporary Debates in Risk Management*, UCL Press, London.

Horlick-Jones, T. (1996) The problem of blame. In Hood, C. and Jones, D. K. C. (eds) *Accident and Design: Contemporary Debates in Risk Management*, UCL Press, London, 61–70.

Hughes, W. P., Hillebrandt, P. and Murdoch, J. R. (2000) The impact of contract duration on the cost of cash retention, *Construction Management and Economics*, 18 (1), 11–14.

Ibbs, C. W. and Ashley, D. B. (1987) Impact of various construction contract clauses, *Journal of Construction Engineering and Management*, 113 (3), 501–521.

ICE (1998) *RAMP*, The Institute of Civil Engineers and the Faculty and Institute of Actuaries, Thomas Telford, London.

Janis, I. L. (1988) Groupthink. In Lau, J. B. and Shani, A. B. (eds) *Behavior in Organizations – An Experiential Approach*, Fourth Edition, Irwin Homewood, Urbana, Illinois, 162–169.

Jarman, A. and Kouzmin, A. (1990) Decision pathways from crisis – A contingency theory simulation heuristic for the challenger space disaster (1983–1988). In Block, A. (ed.) *Contemporary Crisis – Law, Crime and Social Policy*, Kluwer Academic Press, Netherlands, 399–433.

Kasperson, R. and Kasperson, J. (1996) The social amplification and attenustion of risk, *Journal of Environmental Risk Planning and Management*, May, 64–77.

King, N. (1992) Modelling the innovation process: An empirical comparison of approaches, *Journal of Occupational and organisational Psychology*, Wiley, Chichester.

King, N. and Anderson, N. (1995) *Innovation and Change in Organizations*, Routledge, London.

Klein, R. (1997) Questions for cash, *Building*, 11 April, 38–39.

Klein, R. (2000), The politics of risk: The case of BSE, *British Medical Journal*, London, November, 321 (7269), 1091–1092.

Kumeraswamy, M. M., Palaneeswaran, E. and Humphreys, P. (2000) Selection matters – in construction chain optimisation, *International Journal of Physical Distribution in Logistics Management*, 30 (7/8), 661–680.

Kutner, M. (1996) Coping with crisis, *Occupational Health and Safety*, 65 (2), 22–24.

LaPlante, A. (1998) 90s style brainstorming. In Davis, M. H. (ed.) *Social Psychology*, Dushkin Connecticut, Guilford, USA.

Latham, M. Sir (1994) *Constructing the Team*, HMSO, London.

Lawson, B. (1990) *How Designers Think*, Second Edition, Butterworth Architecture, Cambridge.

Lawson, B. R. (1979) Cognitive studies in architectural design, *Ergonomics*, 22 (1), 59–68.

Lerbinger, O. (1997) *The Crisis Manager: Facing Risk and Responsibility*, Lawrence Erlbaum Associates Publishers, New Jersey, USA.

Loosemore, M. (2000) *Crisis Management in Construction Projects*, American Society of Civil Engineers Press, New York, USA.

Loosemore, M. and Lam, A. (2004) Opportunity management; the locus of control in health and safety, *Construction Management and Economics*, 22 (4), 385–394.

Loosemore, M. and McGeorge, D. (2002) *An International Comparison of Reform Agendas in the Building and Construction Industry* – Discussion Paper Six, Royal Commission into the Australian Building and Construction Industry, Melbourne, Victoria.

Loosemore, M. and Teo, M. (2002) The crisis preparedness of construction companies, *ASCE Journal of Management in Engineering*, 16 (5), 51–60.

Loosemore, M., Lingard, H., Walker, D. H. T. and MacKenzie, J. (1999) Benchmarking safety management systems in contracting organisations against best practice in other industries. In Singh, A., Hinze, J. and Coble, R. J. (eds) *Implementation of Safety and Health on Construction Sites*, CIB W99 Second International Conference, Honolulu, Hawaii, 883–889.

Loosemore, M., Nguyen, B. T. and Dennis, N. (2000) Encouraging conflict in the construction industry, *Construction Management and Economics*, 18 (4), 447–457.

Loosemore, M., Dainty, A. and Lingard, H. (2004) *Strategic Human Resource Management in the Construction Industry*, Taylor & Francis, London, UK.

Madine, V. (2002) Catcher in the team, *Building*, 24 May, 26–28.

Mindszenthy, B. J., Watson, T. A. G. and Koch, W. J. (1988) *No Surprises: The Crisis Communications Management System*, Bedford House Publishing Ltd, Ontario, Canada.

Mitroff, I. and Pearson, C. (1993) *Crisis Management: A Diagnostic Guide for Improving Your Organisation's Crisis Preparedness*, Jossey-Bass Publishers, San Francisco, USA.

Moodley, K. and Preece, C. N. (1996) Implementing community policies in the construction industry. In Langford, D. A. and Retik, A. (eds) *The Organization and Management of Construction*, E & FN Spon, London, 178–186.

More, E. (1995) Crisis management and communication in Australian organisations, *Australian Journal of Communication*, 22 (1), 31–47.

Morgan, M. G. and Keith, D. W. (1995) Subjective judgements by climate experts, *Environmental Science and Technology*, 29, 468–476.

Murdoch, J. and Hughes, W. (1996) *Construction Contracts – Law and Management*, Second Edition, E & FN Spon, London.

National Occupational Health and Safety Commission (2002) *The Role of Design Issues in Work Related Injuries in Australia – 1997–2002*, National Occupational Health and Safety Commission, Canberra, Australia.

Nicodemus, J. (1997) Operational crisis management, *Secured Lender*, 53 (6), 84–90.

Niemeyer, M. V. (1998) Managing risk on design-build projects: The surety's perspective, *Rough Notes*, Indianapolis, March, 141 (3), 56–64.

Nonaka. I. (1991) The knowledge creating company, *Harvard Business Review*, November–December, 96–105.

Nystrom, H. (1990) Organisational innovation. In West, M. A. and Farr, J. L. (eds) *Innovation and Creativity at Work: Psychological and Organisational Strategies*, Wiley, Chichester.

Odeyinka, H. A. (2000) An evaluation of the use of insurance in managing construction risks, *Construction Management and Economics*, 18 (5), 519–525.

Osborn, A. (1953) *Applied Imagination*, Scribner's, New York.

Parisotti, M. (2001) What apocalypse? *Building*, 10 August, 37.

Parnes, S. J., Noller, R. B. and Biondi, A. M. (1977) *Guide to Creative Action*, Scribner, New York.

Pascale, R. T. (1991) *Managing on the Edge*, Penguin Books, Harmondsworth, UK.

Pearson, C. M. and Clair, J. A. (1998) Reframing crisis management, *Academy of Management Review*, 23 (1), 59–76.

Pearson, C. M., Misra, S. K., Clair, J. A. and Mitroff, I. (1997) Managing the unthinkable, *Organisational Dynamics*, Autumn, 26 (2), 51–64.

Perinotto, T. (2002) Designing for a post-September 11 world, *Australian Financial Review News*, 5 April, 45.

Philips, J. J. (2001) *HR Trends Worldwide*, Gulf Publishing Company, Houston Texas.

Pierre, St, R., Herendeem, N. M., Moore, D. S. and Nagle, A. M. (1996) Does occupational stereotyping still exist? *The Journal of Personality and Social Psychology*, 39 (5), 832–845.

PMI (1992) *Project and Program Risk Management – A Guide to Managing Project Risks and Opportunities*, Project Management Institute, Newtown Square, Pennsylvania, USA.

PMI (2002), *Risk Management Maturity Level Development*, Project Management Institute, London, UK.

Popper, K. (1959) *The Logic of Scientific Discovery*, Hutchinson, London.

Preece, C. N., Moodley, K. and Smith, A. M. (1998) *Corporate Communications in Construction*, Blackwell Science, Oxford, UK.

Richardson, B. (1996) Modern management's role in the demise of a sustainable society, *Journal of Contingencies and Crisis Management*, March, 4 (1), 20–31.

RICS (2003) *Collateral Warranties*, Briefing papers ref: confac/031, RICS, May.

Royer, P. S. (2000) Risk management: The undiscovered dimension of project management, *Project Management Journal*, March, 31 (1), 6–13.

Ryan, K. D. and Oestreich, D. K. (1998) *Driving Fear Out of the Workplace*, Second Edition, Jossey-Bass Publishers, San Francisco.

Sagan, S. D. (1991) Rules of engagement. In George, A. L. (ed.) *Avoiding War – Problems of Crisis Management*, Westview Press, San Francisco, 443–470.

Salazar J. G. (2000) Damming the child of ocean: The three gorges project, *Journal of Environment and Development*, 9 (2), 160–174.

Santrock, J. W. (1998) *Psychology*, Fifth Edition, McGraw-Hill, Dubuque, Canada.

Schenk, C. R. (1984), Devils advocacy in managerial decision-making, *Journal of Management Studies*, 21 (2), 153–168.

Sharpe, W. (2004) Talking points: Managing stakeholder relations in PPP projects, *Public Infrastructure Bulletin*, March, 8–15.

Sheaffer, Z., Richardson, B. and Rosenblatt, Z. (1998) Early warning signals management: A lesson from the bearings crisis, *Journal of Contingencies and Crisis Management*, 6 (1), 1–23.

Shen, L. Y. (1999) Risk management. In Best, R. and De Valence, G. (eds) *Building in Value*, Arnold, London, 248–267.

Sikich, G. W. (1993) *It Can't Happen Here: All Hazards Crisis Management Planning*, PennWell Publishing Company, Oklahoma, USA.

Slovic, P., Fischhoff, B. and Lichtenstein, S. (1981) *Perceived Risk: Psychological Factors and Social Implications*, Proceedings of the Royal Society, London, A 376, 17–34.

Smee, R. (2002) *Construction Risk*, Presentation Services, London, UK.

Stebbings, S. (2000) Satisfaction guarantees, *Building*, 13 October, 84.

Stewart, B. (2004) A risky business, *Director*, Australian Institute of Company Directors, 20 (5), June, 31–33.

Stewart, R. W. and Fortune, J. (1995) Applications of systems thinking to the identification, avoidance and prevention of risk, *International Journal of Project Management*, 13 (5), 279–286.

Szilagyi, A. and Wallace, M. J. (1987) *Organisational Behaviour and Performance*, Fourth Edition, Scott, Forestman and Company, Illinois.

Taylor, M. (2000) *Avoiding Claims in Building Design*, Blackwell Science, London,

Teo, M. M. M. (2001) *Operatives Attitudes Towards Waste on Construction Projects*, Unpublished MSc Thesis, University of New South Wales, Sydney, Australia.

The Economist (2004) Just how rotten? *The Economist, www.economist.co.uk/business/displayStory.cfm?story_id=...*, 21 October, London, UK.

Timson, L. (2003) Flirting with disaster, *The Sydney Morning Herald*, Tuesday, 14 October, Cover Story, 5–6.

Trench, D. (1991) *On Target – Design and Manage Cost Procurement System*, Thomas Telford, London.

Turner, B. A. and Pigeon, N. F. (1997) *Man-made Disasters*, Second Edition, Butterworth Heinemann, Oxford, UK.

Tversky, A. and Kahneman, D. (1981) The framing of decisions and the psychology of choice, *Science*, 211, 453–458.

Tversky, A. and Koehler, D. J. (1994) Support theory: A non-extensional representation of subjective probability, *Psychological Review*, 101, 547–567.

Uff, J. (1995) Contract documents and the division of risk. In Uff, J. and Odams, A. M. (eds) *Risk Management and Procurement in Construction*, Centre for Construction Law and Management, Kings College, London, 49–69.

US Government (1976) *Teton Dam Disaster*, Committee on Government Operations, Washington, DC, USA.

Walker, C. and Smith, A. J. (1995) *Privatised Infrastructure: The BOT Approach*, Thomas Telford, London.

Wallach, M. A. (1985) Creativity testing and giftedness. In Horowitz, F. D. and O'Brian, M. (eds) *The Gifted and the Talented: Developmental Perspectives*, American Psychological Association, Washington, DC, USA.

West, M. (2004) Real good kick in the big end of town, *The Australian*, Wednesday, 22 September, 4, Sydney, Australia.

Wilson, R. and Crouch, E. A. C. (2001) *Risk-Benefit Analysis*, Harvard University Press, Harvard, USA.

WorkCover (2001) *Occupational Health and Safety Regulation 2001*, WorkCover New South Wales, Sydney, Australia.

Yeo, K. T. (1990) Risks, classification of estimates, and contingency management, *ASCE Journal of Management in Engineering*, 6 (4), 458–470.

Index